SCIENCE AT THE EDGE

*Conversations with
the Leading Scientific
Thinkers of Today*

Books by John Brockman

As Author:
By the Late John Brockman
37
Afterwords
The Third Culture
Digerati

As Editor:
About Bateson
Speculations
Doing Science
Ways of Knowing
Creativity
The Greatest Inventions of the Past 2,000 Years
The Next Fifty Years
Curious Minds
What We Believe But Cannot Prove
My Einstein
Intelligent Thought
What Is Your Dangerous Idea?
What Are You Optimistic About?

As Coeditor:
How Things Are

SCIENCE AT THE EDGE

*Conversations with
the Leading Scientific
Thinkers of Today*

edited by
JOHN BROCKMAN

UNION SQUARE PRESS
An imprint of Sterling Publishing Co., Inc.

New York / London
www.sterlingpublishing.com

STERLING and the distinctive Sterling logo are
registered trademarks of Sterling Publishing Co., Inc.

Library of Congress Cataloging-in-Publication Data Available

2 4 6 8 10 9 7 5 3 1

Published by Sterling Publishing Co., Inc.
387 Park Avenue South, New York, NY 10016
© 2008 by John Brockman
Previously published by Barnes and Noble in 2003 under the title
The New Humanists: Science at the Edge.
Distributed in Canada by Sterling Publishing
ᶜ/o Canadian Manda Group, 165 Dufferin Street
Toronto, Ontario, Canada M6K 3H6
Distributed in the United Kingdom by GMC Distribution Services
Castle Place, 166 High Street, Lewes, East Sussex, England BN7 1XU
Distributed in Australia by Capricorn Link (Australia) Pty. Ltd.
P.O. Box 704, Windsor, NSW 2756, Australia

Manufactured in the United States of America
All Rights Reserved

Sterling ISBN-13: 978-1-4027-5450-0
ISBN-10: 1-4027-5450-7

For information about custom editions, special sales, premium and
corporate purchases, please contact Sterling Special Sales
Department at 800-805-5489 or specialsales@sterlingpublishing.com.

TABLE OF CONTENTS

INTRODUCTION:
THE EXPANDING THIRD CULTURE

JOHN BROCKMAN

Many people, even many scientists, have a narrow view of science as controlled, replicated experiments performed in the laboratory—and as consisting quintessentially of physics, chemistry, and molecular biology. The essence of science is conveyed by its Latin etymology: scientia, meaning knowledge. The scientific method is simply that body of practices best suited for obtaining reliable knowledge. The practices vary among fields: the controlled laboratory experiment is possible in molecular biology, physics, and chemistry, but it is either impossible, immoral, or illegal in many other fields customarily considered sciences, including all of the historical sciences—astronomy, epidemiology, evolutionary biology, most of the earth sciences, and paleontology. If the scientific method can be defined as those practices best suited for obtaining knowledge in a particular field, then science itself is simply the body of knowledge obtained by those practices.

Just as science—that is, reliable methods for obtaining knowledge—has encroached on areas formerly considered to belong to the humanities (such as psychology), science is also encroaching on the social sciences, especially economics, geography, history, and political science. Not just the broad observation-based and statistical methods of the historical sciences but also detailed techniques of the conventional sciences (such as genetics and molecular biology and animal behavior) are proving essential for tackling problems in the social sciences. Science is the most accurate way of gaining knowledge about anything, whether it is the human spirit, the role of great men in history, or the structure of DNA. Humanities scholars and historians who spurn it condemn themselves to second-rate status and produce unreliable results.

But this doesn't have to be the case. As I wrote in 1991 ("The Emerging Third Culture"): "The third culture consists of those scientists and other thinkers in the empirical world who, through their work and expository writing, are taking the place of the traditional intellectual in rendering visible the deeper meanings of our lives, redefining who and what we are."

There are encouraging signs that the third culture now includes scholars in the humanities who think the way scientists do. They believe that there is a real world and that their job is to understand it and explain it. They test their ideas in terms of logical coherence, explanatory power, and conformity with empirical facts. They do not defer to intellectual authorities: anyone's ideas can be challenged, and understanding progresses and knowledge accumulates through such challenges. They are not reducing the humanities to biological and physical principles, but they do believe that art, literature, history, politics—a whole panoply of humanist concerns—need to take the sciences into account.

Connections do exist: our arts, our philosophies, our literature are the product of human minds interacting with one another, and the human mind is a product of the human brain, which is organized in part by the human genome and has evolved by the physical processes of evolution.

Like scientists, the science-based humanities scholars are intellectually eclectic, seeking ideas from a variety of sources and adopting the ones that prove their worth, rather than working within "systems" or "schools." As such, they are not Marxist scholars, or Freudian scholars, or Catholic scholars. They think like scientists, know science, and easily communicate with scientists; their principal difference from scientists is in the subject matter they write about, not their intellectual style. Science and science-based thinking among enlightened humanities scholars are now part of public culture.

And this is not a one-way street. Just as the science-based humanities scholars are learning from and are influenced by science, scientists are gaining a broader understanding of the import of their own work through interactions with artists.

Something radically new is in the air: new ways of understanding physical systems, new ways of thinking about thinking that call into question many of our basic assumptions. A realistic biology of the mind, advances in physics, electricity, genetics, neurobiology, engineering, the chemistry of materials—all are challenging basic assumptions of who and what we are, of what it means to be human. The arts and the sciences are again joining together as one culture: the third culture.

But evidently this information hasn't caught up to the best and brightest at the most highly regarded newspapers and magazines. Rather than trusting scientists to review books by scientists, publications such as the *New York Times Book Review* and the *New Yorker* turn to literary critics. Confronted with books and ideas that that upend the Freud, Marx, and modernism default, they pussyfoot around the challenge of presenting the public with an accurate representation of knowledge. Why learn about the human genome when you've already read Virginia Woolf?

Not all intellectuals are of this frame of mind. One well-known European intellectual, a distinguished novelist as well as a publisher of literary novels *and* books by physicists, threw up his hands as he exclaimed, "They don't know; they just don't know." To which might be added that a blissful state of ignorance is considered a credential in this world. Why else would reputable publications allow reviewers, ignorant in the sciences, to write about books by scientists?

What can we do about this situation? We can start by asking a question. In 1971, the artist James Lee Byars presented a conceptual piece entitled *The World Question Center*, in which he suggested that to arrive at an axiology of the world's knowledge, it was not necessary to read the 6 million volumes in the Widener Library. His approach was to seek out the most complex and sophisticated minds, put them in a room together, and have them ask each other the questions they are asking themselves.

Here is my question, the question I am asking myself, a question we can ask each other: "Why does society benefit from an accurate representation of knowledge?" For literary intellectuals, asking the question may inspire worthwhile exploration; for scientists and the science minded, it is a question that each can ask in his or her own way and in such a manner that their interrogation of reality may help illuminate the empirical work in their own fields toward the goal of enhancing public knowledge and awareness.

Science at the Edge is an exploration of this new intellectual landscape, in which I track the revolutionary work and ideas of key thinkers in various fields—such as computer science, cosmology, cognition, and evolutionary biology—who are arguing with each other, learning from each other, and applying what they are learning in innovative ways. They are evolutionary biologist Helena Cronin; philosopher Daniel C. Dennett; biogeographer Jared Diamond; technologist

Ray Kurzweil; biological anthropologist Richard Wrangham; computer scientists Rodney Brooks, David Gelernter, Jaron Lanier, Marvin Minsky, Hans Moravec, and Jordan B. Pollack; cognitive scientists Andy Clark and Marc D. Hauser; psychologists Stephen M. Kosslyn and Steven Pinker; and physicists David Deutsch, Alan Guth, Seth Lloyd, Lisa Randall, Martin Rees, Lee Smolin, and Paul Steinhardt. *Science at the Edge* attempts to render visible a revolution from the inside, since the debates that surface here will define the coming decades of scientific thought.

The choice of scientists included in this book is, obviously, far from comprehensive. Some of them I work with professionally: they are clients of my literary agency. Others are not. (Indeed, the great percentage of scientists I represent are not included here.) The selection is serendipitous and has much to do with my personal scientific interests. Most of the chapters are based on interviews I conducted; the rest—essays by David Gelernter, Hans Moravec, Jaron Lanier, Andy Clark, and Jared Diamond—have all been published on *Edge* (www.edge.org), a Web site I launched in 1997 that is devoted to discussions among scientists at the frontiers of their disciplines.

The origin of the *Edge* community is an informal assembly of scientists and other empirical thinkers known as the Reality Club, which I put together in the early 1980s. The club's members were individuals accustomed to creating their own reality and rejecting an ersatz, appropriated one; they were (and are) people out there doing it rather than talking about it. In the beginning, the Reality Club held its meetings in Chinese restaurants, artists' lofts, museums, living rooms, and the board rooms of Rockefeller University, the New York Academy of Sciences, and various investment banking firms, among other venues. *Edge*, a nonprofit foundation established in 1888, is the offspring of the Reality Club. It has now migrated to the Internet. Here you will find a number of today's sharpest minds taking their ideas into the bullring, in the full expectation that these ideas will be challenged. *New Scientist* has called the site "breathtaking in scope" and has hailed it for asking "big, deep and ambitious questions—questions that suggest that science is finally edging into the domain of philosophy and religion."

A few *Edge* contributors are bestselling authors or otherwise well known in the mass culture. Most are not. *Edge* encourages work on the perimeters of our culture and the investigation of ideas that have not been generally exposed, hence the motto of the community: "Arrive at the edge of the world's knowledge, seek out the most complex and sophisticated minds, put them in a room together, and have them ask each other the questions they are asking themselves." *Edge* is a point of view, not just a group of people. Its contributors share with each other the boundaries of their knowledge and respond vigorously to the comments, criticisms, and insights of their peers. *Wired* magazine once described *Edge* as "A-list: . . . recreates Dorothy Parker's Vicious Circle without the food and alcohol. . . . A brilliant format, partly because of who's on the list—Richard Dawkins, Freeman Dyson, David Gelernter, Nathan Myhrvold, and Naomi Wolf, to name a few." But *Edge* is an altogether different group from such gatherings as the Algonquin Round Table, the Apostles, or the Bloomsbury Group. It does, however, offer the same quality of intellectual adventure. Perhaps the closest resemblance is to the eighteenth-century Lunar Society of Birmingham, an informal club of the leading cultural figures of the coming industrial age—James Watt, Erasmus Darwin, Josiah Wedgwood, Joseph Priestley, Matthew Boulton, and William

Withering, among others. In a similar fashion, the *Edge* community brings together those who are exploring the themes of the postindustrial age. *Edge* has featured a wide range of people in the arts and sciences: cultural anthropologist Mary Catherine Bateson on bridging cultural gaps, evolutionary biologist Richard Dawkins on the public's view of science, physicist Freeman Dyson on the ultimate future of life in the universe, musician Brian Eno on the creation of cultural values, psychologist Howard Gardner on educational reform, biologist Stuart Kauffman on time in quantum cosmology, and psychologist Judith Rich Harris on how personality is formed.

In the interviews and conversations presented here, I have taken the editorial license to reproduce my tapes in essay form. Having assumed that the views of *Edge's* participants will be of far more interest to readers than my own ideas on their areas of expertise, I have written myself (and my questions) out of the text. But although the interviewees have read, and in some cases edited, the transcriptions of their spoken words, there is no intention that these chapters in any way represent their own writing. For that, read their books, listed in the Suggested Reading appendix.

When my essay on "The New Humanists" appeared on *Edge* in April 2002, it brought a record number of responses—including the occasional impassioned refutation—from members of the *Edge* mailing list. The epilogue contains a sampling of this pungent commentary from some of the New Humanists themselves.

John Brockman
New York, December 2007

PART I
HOMO SAPIENS

THE NEW HUMANISTS

JOHN BROCKMAN

In 1991, in an essay entitled "The Emerging Third Culture," I put forward the following argument:

> In the past few years, the playing field of American intellectual life has shifted, and the traditional intellectual has become increasingly marginalized. A 1950s education in Freud, Marx, and modernism is not a sufficient qualification for a thinking person today. Indeed, the traditional American intellectuals are, in a sense, increasingly reactionary, and quite often proudly (and perversely) ignorant of many of the truly significant intellectual accomplishments of our time. Their culture, which dismisses science, is often non-empirical. It uses its own jargon and washes its own laundry. It is chiefly characterized by comment on comments, the swelling spiral of commentary eventually reaching the point where the real world gets lost.

Twelve years later, that fossil culture has been essentially replaced by the "third culture" of the essay's title—a reference to C. P. Snow's celebrated division of the thinking world into two cultures, that of the literary intellectual and that of the scientist. This new culture consists of those scientists and other thinkers in the empirical world who, through their work and expository writing, have taken the place of the traditional intellectual in rendering visible the deeper meanings of our lives, redefining who and what we are.

The scientists of the third culture share their work and ideas not just with each other but with a newly educated public, through their books. Focusing on the real world, they have led us into one of the most dazzling periods of intellectual activity in human history. The achievements of the third culture are not the marginal disputes of a quarrelsome mandarin class; they affect the lives of everybody on the planet. The emergence of this new culture is evidence of a great intellectual hunger, a desire for the new and important ideas that drive our times: revolutionary developments in molecular biology, genetic engineering, nanotechnology, artificial intelligence, artificial life, chaos theory, massive parallelism, neural nets, the inflationary universe, fractals, complex adaptive systems, linguistics, superstrings, biodiversity, the human genome, expert systems, punctuated equilibrium, cellular automata, fuzzy logic, virtual reality, cyberspace, and teraflop machines. Among others.

HUMANISM AND
THE INTELLECTUAL WHOLE

Around the fifteenth century, the word "humanism" was tied in with the idea of one intellectual whole. A Florentine nobleman knew that to read Dante but ignore science was ridiculous. Leonardo was a great artist, a great scientist, a great technologist. Michelangelo was an even greater artist and engineer. These men were intellectually holistic giants. To them, the idea of embracing humanism while remaining ignorant of the latest scientific and technological achievements would have been incomprehensible. The time has come to reestablish that holistic definition.

In the twentieth century, a period of great scientific advancement, instead of having science and technology at the center of the intellectual world—of having a unity in which scholarship included science and technology along with literature and art—the official culture kicked them out. Traditional humanities scholars looked at science and technology as some sort of technical special product. Elite universities nudged science out of the liberal arts undergraduate curriculum—and out of the minds of many young people, who, as the new academic establishment, so marginalized themselves that they are no longer within shouting distance of the action.

In too much of academia, intellectual debate tends to center on such matters as who was or was not a Stalinist in 1937, or what the sleeping arrangements were for guests at a Bloomsbury weekend in the early part of the twentieth century. This is not to suggest that studying history is a waste of time: History illuminates our origins and keeps us from reinventing the wheel. But the question arises: History of what? Do we want the center of culture to be based on a closed system, a process of text in/text out, and no empirical contact with the real world? One can only marvel at, for example, art critics who know nothing about visual perception; "social constructionist" literary critics uninterested in the human universals documented by anthropologists; opponents of genetically modified foods, additives, and pesticide residues who are ignorant of genetics and evolutionary biology.

CULTURAL PESSIMISM VS. SCIENTIFIC OPTIMISM

A fundamental distinction exists between the literature of science and that of disciplines whose subjects are self-referential and most often concerned with the exegesis of earlier thinkers. Unlike those disciplines in which there is no expectation of systematic progress and in which one reflects on and recycles the ideas of others, science, on its frontiers, poses more and better questions, better put. They are questions phrased to elicit answers; science finds the answers and moves on. Meanwhile, the traditional humanities establishment continues its exhaustive insular hermeneutics, indulging itself in cultural pessimism, clinging to its fashionably glum outlook on world events.

"We live in an era in which pessimism has become the norm," writes Arthur Herman, in *The Idea of Decline in Western History*. Herman, who coordinates the Western Civilization Program at the Smithsonian, argues that the decline of the West, with its view of our "sick society," has become the dominant theme in intellectual discourse, to the point where the very idea of civilization has changed. He continues:

> This new order might take the shape of the Unabomber's radical environmental utopia. It might also be Nietzsche's Overman, or Hitler's Aryan National Socialism, or Marcuse's utopian union of technology and Eros, or Frantz Fanon's revolutionary *fellahin.* Its carriers might be the ecologist's "friends of the earth," or the multiculturalist's "persons of color," or the radical feminist's New Amazons, or Robert Bly's New Men. The particular shape of the new order will vary according to taste; however, its most important virtue will be its totally non-, or even anti-Western character. In the end, what matters to the cultural pessimist is less what is going to be created than what is going to be destroyed—namely, our "sick" modern society. . . . [T]he sowing of despair and self doubt has become so

pervasive that we accept it as a normal intellectual stance—even when it is directly contradicted by our own reality.

Key to this cultural pessimism is a belief in the myth of the Noble Savage—that before we had science and technology, people lived in ecological harmony and bliss. Quite the opposite is the case. That the greatest change continues to be the rate of change must be hard to deal with, if you're still looking at the world through the eyes of Spengler and Nietzsche. In their almost religious devotion to a pessimistic worldview, the academic humanists have created a culture of previous "isms" that turn on themselves and endlessly cycle. How many times have you seen the name of an academic humanist icon in a newspaper or magazine article and immediately stopped reading? You know what's coming. Why waste the time?

As a counternarrative to this cultural pessimism, consider the twofold optimism of science.

First, the more science you do, the more there is to do. Scientists are constantly acquiring and processing new information. This is the reality of Moore's Law—just as there has been a doubling of computer processing power every eighteen months for the past twenty years, so too do scientists acquire information exponentially. They can't help but be optimistic.

And second, much of the new information is either good news or news that can be made good thanks to ever deepening knowledge and ever more efficient and powerful tools and techniques.

Scientists debate continually, and reality is the check. They may have egos as large as those possessed by the iconic figures of the academic humanities, but they handle their hubris in a very different way. They can be moved by arguments, because they work in an empirical world of facts, a world based on reality. There are no fixed, unalterable positions. They are both the creators and the critics of their shared enterprise. Ideas come from them and they also criticize one another's ideas. Through the process of creativity and criticism and debates, they decide which ideas get weeded out and which become part of the consensus that leads to the next level of discovery. Unlike the humanities academicians, who talk about each other, scientists talk about the universe. Moreover, there's not much difference between the style of thinking of a cosmologist trying to understand the physical world by studying the origins of atoms, stars, and galaxies and an evolutionary biologist trying to understand the emergence of complex systems from simple beginnings or trying to see patterns in nature. As exercises, these entail the same mixture of observation, theoretical modeling, computer simulation, and so on—as in most other scientific fields. The worlds of science are convergent. The frame of reference is shared across their disciplines.

Science is still near the beginning. As the frontiers advance, the horizon gets wider and comes into focus. And these advances have changed the way we see our place in nature. The idea that we are an integral part of this universe—a universe governed by physical and mathematical laws that our brains are attuned to understand—causes us to see our place in the unfolding of natural history differently. We have come to realize, through developments in astronomy and cosmology, that we are still quite near the beginning. The history of creation has been enormously expanded—from 6,000 years back to the 13.7 billion years of Big Bang cosmology. But the future has expanded even more—perhaps to infinity. In the seventeenth century, people not only believed in that constricted past but thought that history was near its end: The apocalypse was coming. A realization that time may well be endless leads us to a new view of the human species—as not being in any sense the culmination but perhaps a fairly early stage of the process of evolution. We arrive at this concept through detailed observation and analysis, through science-based thinking; it allows us to see life playing an ever greater role in the future of the universe.

There are encouraging signs that the third culture now includes scholars in the humanities who think the way scientists do. Like their colleagues in the sciences, they believe there is a real world and their job is to understand it and explain it. They test their ideas in terms of logical coherence, explanatory power, conformity with empirical facts. They do not defer to intellectual authorities: Anyone's ideas can be challenged, and understanding and knowledge accumulate through such challenges. They are not reducing the humanities to biological and physical principles, but they do believe that art, literature, history, politics—a whole panoply of humanist concerns—need to take the sciences into account.

Connections do exist: Our arts, our philosophies, our literature are the product of human minds interacting with one another, and the human mind is a product of the human brain, which is organized in part by the human genome and evolved by the physical processes of evolution. Like scientists, the science-based humanities scholars are intellectually eclectic, seeking ideas from a variety of sources and adopting the ones that prove their worth, rather than working within "systems" or "schools." As such, they are not Marxist scholars or Freudian scholars or Catholic scholars. They think like scientists, know science, and easily communicate with scientists; their principal difference from scientists is in the subject matter they write about, not their intellectual style.

Science-based thinking among enlightened humanities scholars is now part of public culture.

In short, something radically new is in the air: new ways of understanding physical systems, new ways of thinking about thinking that call into question many of our basic assumptions. A realistic biology of the mind, advances in physics, information technology, genetics, neurobiology, engineering, the chemistry of materials—all are challenging basic assumptions of who and what we are, of what it means to be human. The arts and the sciences are again joining together as one culture, the third culture. Those involved in this effort—on either side of C. P. Snow's old divide—are at the center of today's intellectual action. They are the new humanists.

A NEW SCIENTIFIC SYNTHESIS
OF HUMAN HISTORY

JARED DIAMOND

Why did human development proceed at such different rates on different continents for the last 13,000 years? . . . Historians tend to avoid this subject like the plague, because of its apparently racist overtones. Many people, or even most people, assume that the answer involves biological differences in average IQ among the world's populations, despite the fact that there is no evidence for the existence of such IQ differences. . . . In case the stink of racism still makes you feel uncomfortable about exploring this subject, just reflect on the underlying reason that so many people accept racist explanations of history's broad pattern: We don't have a convincing alternative explanation. Until we do, people will continue to gravitate by default to racist theories. That leaves us with a huge moral gap, which constitutes the strongest reason for tackling this uncomfortable subject.

JARED DIAMOND is a professor of geography at UCLA, a MacArthur Fellow, winner of the National Medal of Science, and author of *The Third Chimpanzee* (awarded the British Science Book Prize and a *Los Angeles Times* Book Prize), *Collapse,* and the Pulitzer Prize-winning *Guns, Germs, and Steel* and *Collapse.*

I've set myself the modest task of trying to explain the broad pattern of human history on all the continents for the last 13,000 years. Why did history take such different evolutionary courses for peoples of different continents? This problem has fascinated me for a long time, but it's now ripe for a new synthesis because of recent advances in many fields seemingly remote from history, including molecular biology, plant and animal genetics, biogeography, archaeology, and linguistics.

As we all know, Eurasians, especially peoples of Europe and Eastern Asia, have spread around the globe, to dominate the modern world in wealth and power. Other peoples, including most Africans, survived and have thrown off European domination but remain behind in wealth and power. Still other peoples, including the original inhabitants of Australia, the Americas, and Southern Africa, are no longer even masters of their own lands but have been decimated, subjugated, or exterminated by European colonialists. Why did history turn out that way, instead of the opposite way? Why weren't native Americans, Africans, and aboriginal Australians the ones who conquered or exterminated Europeans and Asians?

This big question can easily be pushed back one step further. By the year A.D. 1500, the approximate year when Europe's overseas expansion was just beginning, peoples of the different continents already differed greatly in technology and political organization. Much of Eurasia and North Africa was occupied then by Iron Age states and empires, some of them on the verge of industrialization. Two native American peoples, the Incas and Aztecs, ruled over empires with stone tools and were just starting to experiment with bronze. Parts of sub-Saharan Africa were divided among small indigenous Iron Age states or chiefdoms. But all peoples of Australia, New Guinea, and the Pacific islands, and many peoples of the Americas and sub-Saharan Africa, were still living as farmers or even still as hunter/gatherers with stone tools.

Obviously, those differences as of A.D. 1500 were the immediate cause of the modern world's inequalities. Empires with iron tools conquered or exterminated tribes with stone tools. But how did the world evolve to be the way it was in the year A.D. 1500?

This question, too, can be easily pushed back a further step, with the help of written histories and archaeological discoveries. Until the end of the last Ice Age, around 11,000 B.C., all humans on all continents were still living as Stone Age hunter/gatherers. Different rates of development on different continents from 11,000 B.C. to A.D. 1500 were what produced the inequalities of A.D. 1500. While aboriginal Australians and many native American peoples remained Stone Age hunter/gatherers, most Eurasian peoples and many peoples of the Americas and sub-Saharan Africa gradually developed agriculture, herding, metallurgy, and complex political organization. Parts of Eurasia, and one small area of the Americas, developed indigenous writing as well. But each of these new developments appeared earlier in Eurasia than elsewhere.

So we can finally rephrase our question about the evolution of the modern world's inequalities as follows: Why did human development proceed at such different rates on different continents for the last 13,000 years? Those differing rates constitute the broadest pattern of history, the biggest unsolved problem of history, and my subject in this essay.

Historians tend to avoid this subject like the plague, because of its apparently racist overtones. Many people, or even most people, assume that the answer involves biological differences in average IQ among the world's populations, despite the fact that there is no evidence for the existence of such IQ differences. Even to ask the question about why different peoples had different histories strikes some of us as evil, because it appears to be justifying what happened in history. In fact, we study the injustices of history for the same reason that we study genocide, and for the same reason that psychologists study the minds of murderers and rapists—not in order to justify history, genocide, murder, and rape but to understand how those evil things came about and then to use that understanding so as to prevent their happening again. In case the stink of racism still makes you feel uncomfortable about exploring this subject, just reflect on the underlying reason that so many people accept racist explanations of history's broad pattern: We don't have a convincing alternative explanation. Until we do, people will continue to gravitate by default to racist theories. That leaves us with a huge moral gap, which constitutes the strongest reason for tackling this uncomfortable subject.

Let's proceed continent by continent. As our first continental comparison, let's consider the collision of the Old World and the New World that began with Christopher Columbus's voyage in A.D. 1492, because the proximate factors involved in that outcome are well understood. I'll now give you a summary and interpretation of the histories of North America, South America, Europe, and Asia from my perspective as a biogeographer and evolutionary biologist—all that in ten minutes; two minutes per continent. Here we go:

Most of us are familiar with the stories of how a few hundred Spaniards under Cortés and Pizarro overthrew the Aztec and Inca empires. The populations of each of those empires numbered tens of millions. We're also familiar with the gruesome details of how other Europeans conquered other parts of the New World. The result is that Europeans came to settle and dominate most of the New World, while the native American population declined drastically from its level as of A.D. 1492. Why did it happen that way? Why didn't it instead happen that the Emperors Montezuma or Atahuallpa led the Aztecs or Incas to conquer Europe?

The proximate reasons are obvious. Invading Europeans had steel swords, guns, and horses, while native Americans had only stone and wooden weapons and no animals that could be ridden. Those military advantages repeatedly enabled troops of a few dozen mounted Spaniards to defeat Indian armies numbering in the thousands.

Nevertheless, steel swords, guns, and horses weren't the sole proximate factors behind the European conquest of the New World. Infectious diseases introduced with Europeans, like smallpox and measles, spread from one Indian tribe to another, far in advance of Europeans themselves, and killed an estimated 95 percent of the New World's Indian population. Those diseases were endemic in Europe, and Europeans had had time to develop both genetic and immune resistance to them, but Indians initially had no such resistance. The role played by infectious diseases in the European conquest of the New World was duplicated in many other parts of the world, including aboriginal Australia, Southern Africa, and many Pacific islands.

Finally, there is still another set of proximate factors to consider. How is it that Pizarro and Cortés reached the New World at all, before Aztec and Inca conquistadors could reach Europe? That outcome depended partly on technology, in the form of oceangoing ships. Europeans had such ships, while the Aztecs and Incas did not. Also, those European ships were backed by the centralized political organization that enabled Spain and other European countries to build and staff the ships. Equally crucial was the role of European writing in permitting the quick spread of accurate detailed information, including maps, sailing directions, and accounts by earlier explorers, back to Europe, to motivate later explorers.

So far we've identified a series of proximate factors behind European colonization of the New World: namely, ships, political organization, and writing, which brought Europeans to the New World; European germs, which killed most Indians before they could reach the battlefield; and guns, steel swords, and horses, which gave Europeans a big advantage on the battlefield. Now let's try to push the chain of causation back further. Why did these proximate advantages go to the Old World rather than to the New World? Theoretically, native Americans might have been the ones to develop steel swords and guns first, to develop oceangoing ships and empires and writing first, to be mounted on domestic animals more terrifying than horses, and to bear germs worse than smallpox.

The part of that question that's easiest to answer concerns the reasons why Eurasia evolved the nastiest germs. It's striking that native Americans evolved no devastating epidemic diseases to give to Europeans, in return for the many devastating epidemic diseases that Indians received from the Old World. There are two straightforward reasons for this gross imbalance: First, most of our familiar epidemic diseases can sustain themselves only in large, dense human populations concentrated into villages and cities, which arose much earlier in the Old World than in the New. Second, recent studies of microbes by molecular biologists have shown that most human epidemic diseases evolved from similar epidemic diseases of the dense populations of Old World domestic animals, with which we came into close contact. For example, measles and tuberculosis evolved from diseases of our cattle, influenza from a disease of pigs, and smallpox possibly from a disease of camels. The Americas had very few native domesticated animal species from which humans could acquire such diseases.

Let's now push the chain of reasoning back one step further. Why were there far more species of domesticated animals in Eurasia than in the Americas? The Americas harbor over 1,000 native wild mammal species, so you might initially suppose that the Americas offered plenty of starting material for domestication. In fact, only a tiny fraction of wild mammal species has been successfully domesticated, because domestication requires that a wild animal fulfill many prerequisites. The animal has to have a diet that humans can supply, a rapid growth rate, a willingness to breed in captivity, a tractable disposition, a social structure involving submissive behavior toward dominant animals and humans, and lack of a tendency to panic when fenced in. Thousands of years ago, humans domesticated every possible large wild mammal species fulfilling all those criteria and worth domesticating, with the result that there have been no valuable additions of domestic animals in recent times, despite the efforts of modern science.

Eurasia ended up with the greatest number of domesticated animal species in part because it is the world's largest land mass and offered the greatest number of wild species to begin with. That preexisting difference was magnified 13,000 years ago at the end of the last Ice Age, when most of the large mammal species of North and South America became extinct, perhaps exterminated by the first arriving Indians. As a result, native Americans inherited far fewer species of big wild mammals than did Eurasians, leaving them only with the llama and alpaca as a domesticate. Differences between the Old and New Worlds in domesticated plants, especially in large-seeded cereals, are qualitatively similar to these differences in domesticated mammals, though the difference is not so extreme.

Another reason for the higher local diversity of domesticated plants and animals in Eurasia is that Eurasia's main axis is east/west, whereas the main axis of the Americas is north/south. Eurasia's east/west axis meant that species domesticated in one part of Eurasia could easily spread thousands of miles at the same latitude, encountering the same climate and day length to which they were already adapted. As a result, chickens and citrus fruit domesticated in Southeast Asia quickly spread westward to Europe; horses domesticated in the Ukraine quickly spread eastward to China; and the sheep, goats, cattle, wheat, and barley of the Fertile Crescent quickly spread both west and east. In contrast, the north/south axis of the Americas meant that species domesticated in one area couldn't spread far without encountering climates and day lengths to which they were not adapted. As a result, the turkey never spread from its site of domestication in Mexico to the Andes; llamas and alpacas never spread from the Andes to Mexico, so that the Indian civilizations of Central and North America remained entirely without pack animals; and it took thousands of years for the corn that evolved in Mexico's climate to become modified into a corn adapted to the short growing season and seasonally changing day length of North America.

Eurasia's domesticated plants and animals were important for several other reasons besides letting Europeans develop nasty germs. Domesticated plants and animals yield far more calories per acre than do wild habitats, in which most species are inedible to humans. As a result, population densities of farmers and herders are typically 10 to 100 times greater than those of hunter/gatherers. That fact alone explains why farmers and herders everywhere in the world have been able to push hunter/gatherers off of land suitable for farming and herding. Domestic animals revolutionized land transport. They also revolutionized agriculture, by letting a farmer plow and manure much more land than he could by his own efforts. Also, hunter/gatherer societies tend to be egalitarian and to have no political organization beyond the level of the band or tribe, whereas the food surpluses and storage made possible by agriculture permitted the development of stratified, politically centralized societies with governing elites. Those food surpluses also accelerated the development of technology, by supporting craftspeople who didn't raise their own food and who could instead devote themselves to developing metallurgy, writing, swords, and guns.

Thus we began by identifying a series of proximate explanations—guns, germs, and so on—for the conquest of the Americas by Europeans. Those proximate factors seem to me ultimately traceable in large part to the Old World's greater number of domesticated plants, much greater number of domesticated animals, and east/west axis. The chain of causation is most direct in explaining the Old World's advantages of horses and nasty germs. But domesticated plants and animals also led more indirectly to Eurasia's advantage in guns, swords, oceangoing ships, political organization, and writing, all of which were products of the large, dense, sedentary, stratified societies made possible by agriculture.

Let's next examine whether this scheme, derived from the collision of Europeans with native Americans, helps us understand the broadest pattern of African history, which I'll summarize in five minutes. I'll concentrate on the history of sub-Saharan Africa, because it was much more isolated from Eurasia by distance and climate than was North Africa, whose history is closely linked to Eurasia's history. Here we go again:

Just as we asked why Cortés invaded Mexico before Montezuma could invade Europe, we can similarly ask why Europeans colonized sub-Saharan Africa before sub-Saharans could colonize Europe. The proximate factors were the familiar ones of guns, steel, oceangoing ships, political organization, and writing. But again, we can ask why guns and ships and so on ended up being developed in Europe rather than in sub-Saharan Africa. To the student of human evolution, that question is particularly puzzling, because humans have been evolving for millions of years longer in Africa than in Europe, and even anatomically modern *Homo sapiens* may have reached Europe from Africa only within the last 50,000 years. If time were a critical factor in the development of human societies, Africa should have enjoyed an enormous head start and advantage over Europe.

Again, that outcome largely reflects biogeographic differences in the availability of domesticable wild animal and plant species. Taking first domestic animals, it's striking that the sole animal domesticated within sub-Saharan Africa was a bird, the guinea fowl. All of Africa's mammalian domesticates—cattle, sheep, goats, horses, even dogs—entered sub-Saharan Africa from the north, from Eurasia or North Africa. At first that sounds astonishing, since we now think of Africa as the continent of big wild mammals. In fact, none of those famous big wild mammal species of Africa proved domesticable. They were all disqualified by one or another problem, such as unsuitable social organization, intractable behavior, slow growth rate, and so on. Just think what the course of world history might have been if Africa's rhinos and hippos had lent themselves to domestication! If that had been possible, African cavalry mounted on rhinos or hippos would have made mincemeat of European cavalry mounted on horses. But it couldn't happen.

Instead, as I mentioned, the livestock adopted in Africa were Eurasian species that came in from the north. Africa's long axis, like that of the Americas, is north/south rather than east/west. Those Eurasian domestic mammals spread southward very slowly in Africa, because they had to adapt to different climate zones and different animal diseases.

The difficulties posed by a north/south axis to the spread of domesticated species are even more striking for African crops than they are for livestock. Remember that the food staples of ancient Egypt were Fertile Crescent and Mediterranean crops like wheat and barley, which require winter rains and seasonal variation in day length for their germination. Those crops couldn't spread south in Africa past Ethiopia, beyond which the rains come in the summer and there's little or no seasonal variation in day length. Instead, the development of agriculture in the sub-Sahara had to await the domestication of native African plant species, like sorghum and millet, adapted to Central Africa's summer rains and relatively constant day length. Ironically, those crops of Central Africa were for the same reason then unable to spread south to the Mediterranean zone of South Africa, where once again winter rains and big seasonal variations in day length prevailed. The southward advance of native African farmers with Central African crops halted in Natal, beyond which Central African crops couldn't grow—with enormous consequences for the recent history of South Africa.

In short, a north/south axis and a paucity of wild plant and animal species suitable for domestication were decisive in African history, just as they were in native American history. Although native Africans domesticated some plants in the Sahel and Ethiopia and tropical West Africa, they acquired valuable domestic animals only later, from the north. The resulting advantages of Europeans in guns, ships, political organization, and writing permitted Europeans to colonize Africa rather than Africans to colonize Europe.

Let's now conclude our whirlwind tour around the globe by devoting two minutes to the last continent, Australia. Here we go again, for the last time:

In modern times, Australia was the sole continent still inhabited only by hunter/gatherers. That makes Australia a critical test of any theory about continental differences in the evolution of human societies. Native Australia had no farmers or herders, no writing, no metal tools, and no political organization beyond the level of the tribe or band. Those, of course, are the reasons why European guns and germs destroyed aboriginal Australian society. But why had all native Australians remained hunter/gatherers?

There are three obvious reasons. First, even to this day no native Australian animal species and only one plant species (the macadamia nut) have proved suitable for domestication. There still are no domestic kangaroos.

Second, Australia is the smallest continent, and most of it can support only small human populations because of low rainfall and productivity. Hence the total number of Australian hunter/gatherers was only about 300,000.

Finally, Australia is the most isolated continent. The sole outside contacts of aboriginal Australians were tenuous overwater contacts with New Guineans and Indonesians.

To get an idea of the significance of that small population size and isolation for the pace of development in Australia, consider the Australian island of Tasmania, which had the most extraordinary human society in the modern world. Tasmania is an island of modest size, but it was the most extreme outpost of the most extreme continent, and it illuminates a big issue in the evolution of all human societies. Tasmania lies 130 miles southeast of Australia. When it was first visited by Europeans in 1642, Tasmania was occupied by 4,000 hunter/gatherers related to mainland Australians but with the simplest technology of any recent people on Earth. Unlike mainland aboriginal Australians, Tasmanians couldn't start a fire; they had no boomerangs, spear throwers, or shields; they had no bone tools, no specialized stone tools, and no compound tools like an axe head mounted on a handle; they couldn't cut down a tree or hollow out a canoe; they lacked sewing to make sewn clothing, despite Tasmania's cold winter climate with snow; and, incredibly, though they lived mostly on the seacoast, the Tasmanians didn't catch or eat fish. How did those enormous gaps in Tasmanian material culture arise?

The answer stems from the fact that Tasmania used to be joined to the southern Australian mainland at Pleistocene times of low sea level, until that land bridge was severed by a rising sea level 10,000 years ago. People walked out to Tasmania tens of thousands of years ago, when it was still part of Australia. Once that land bridge was severed, though, Tasmanians had absolutely no further contact with mainland Australians or with any other people on Earth until the European arrival in 1642, because both Tasmanians and mainland Australians lacked watercraft capable of crossing those 130-mile straits between Tasmania and Australia. Tasmanian history is thus a study of human isolation unprecedented except in science fiction—namely, complete isolation from other humans for 10,000 years. Tasmania had the smallest and most isolated human population in the world. If population size and isolation have any effect on the accumulation of inventions, we should expect to see that effect in Tasmania.

If all those technologies that I mentioned, absent from Tasmania but present on the opposite Australian mainland, were invented by Australians within the last 10,000 years, we can surely conclude at least that Tasmania's tiny population didn't invent them independently. Astonishingly, the archaeological record demonstrates something further: Tasmanians actually abandoned some technologies they brought with them from Australia and which persisted on the Australian mainland. For example, bone tools and the practice of fishing were both present in Tasmania at the time the land bridge was severed, and both had disappeared from Tasmania by around 1500 B.C. That represents the loss of valuable technologies: Fish could have been smoked to provide a winter food supply, and bone needles could have been used to sew warm clothes.

What sense can we make of these cultural losses?

The only interpretation that makes sense to me goes as follows: First, technology has to be either invented or adopted. Human societies vary in lots of independent factors affecting their openness to innovation. Hence the higher the human population and the more societies there are on an island or continent, the greater the chance of any given invention being conceived and adopted somewhere there.

Second, for all human societies except those of totally isolated Tasmania, most technological innovations diffuse in from the outside instead of being invented locally, so one expects the evolution of technology to proceed most rapidly in societies most closely connected with outside societies.

Finally, technology has to be not only adopted but maintained. All human societies go through fads in which they temporarily either adopt practices of little use or abandon practices of considerable use. Whenever such economically senseless taboos arise in an area with many competing human societies, only some societies will adopt the taboo at a given time. Other societies will retain the useful practice and will either outcompete the societies that lost it or will be there as a model for the societies with the taboos to repent their error and reacquire the practice. If Tasmanians had remained in contact with mainland Australians, they could have rediscovered the value and techniques of fishing and making bone tools that they lost. But that couldn't happen in the complete isolation of Tasmania, where cultural losses became irreversible.

In short, the message of the differences between Tasmanian and mainland Australian societies seems to be the following: All other things being equal, the rate of human invention is faster, and the rate of cultural loss is slower, in areas occupied by many competing societies with many individuals and in contact with societies elsewhere. If this interpretation is correct, then it's likely to be of much broader significance. It probably provides part of the explanation of why native Australians, on the world's smallest and most isolated continent, remained Stone Age hunter/gatherers, while people of other continents were adopting agriculture and metal. It's also likely to contribute to the differences I already discussed between the farmers of sub-Saharan Africa, the farmers of the much larger Americas, and the farmers of the still larger Eurasia.

Naturally, there are many important factors in world history that I haven't had time to elaborate on. For example, I've said little or nothing about the distribution of domesticable plants; about the precise way in which complex political institutions and the development of writing and technology and organized religion depend on agriculture and herding; about the fascinating reasons for the differences within Eurasia between China, India, the Near East, and Europe; and about the effects on history of individuals and of cultural differences unrelated to the environment. But it's now time to summarize the overall meaning of this whirlwind tour through human history, with its unequally distributed guns and germs.

The broadest pattern of history—namely, the differences between human societies on different continents— seems to me to be attributable to differences among continental environments and not to biological differences among peoples themselves. In particular, the availability of wild plant and animal species suitable for domestication and the ease with which those species could spread without encountering unsuitable climates contributed decisively to the varying rates of rise of agriculture and herding; which in turn contributed decisively to the rise of human population numbers, population densities, and food surpluses; which in turn contributed decisively to the development of epidemic infectious diseases, writing, technology, and political organization. In addition, the histories of Tasmania and Australia warn us that the differing areas and isolations of the continents, by determining the number of competing societies, may have been another important factor in human development.

As a biologist practicing laboratory experimental science, I'm aware that some scientists may be inclined to dismiss these historical interpretations as unprovable speculation because they're not founded on replicated laboratory experiments. The same objection can be raised against any of the historical sciences, including astronomy, evolutionary biology, geology, and paleontology. The objection can, of course, be raised against the whole field of history and most of the other social sciences. That's the reason we're uncomfortable about considering history as a science. It's classified as a social science, which is considered not quite scientific. But remember that the word "science" isn't derived from the Latin word for "replicated laboratory experiment" but from the Latin *scientia*, meaning "knowledge." In science, we seek knowledge by whatever methodologies are available and appropriate. There are many fields that no one hesitates to consider sciences, even though replicated laboratory experiments in those fields would be immoral, illegal, or impossible. We can't manipulate some stars while maintaining other stars as controls; we can't start and stop ice ages; and we can't experiment with designing and evolving dinosaurs. Nevertheless, we can still gain considerable insight into those historical fields by other means. Then we should surely be able to understand human history, because introspection and preserved writings give us far more insight into the ways of past humans than we have into the ways of past dinosaurs. For that reason, I'm optimistic that we can eventually arrive at convincing explanations for these broadest patterns of human history.

A BIOLOGICAL UNDERSTANDING OF HUMAN NATURE

STEVEN PINKER

I believe that there is a quasi-religious theory of human nature prevalent among pundits and intellectuals which includes both empirical assumptions about how the mind works and a set of values that people hang on those assumptions. The theory has three parts: [T]he Blank Slate—that we have no inherent talents or temperaments because the mind is shaped completely by the environment (parenting, culture, and society). The second is the myth of the Noble Savage— that evil motives are not inherent in people but spring from corrupting social institutions. The third is the Ghost in the Machine—that the most important part of us is somehow independent of our biology, so that our ability to have experiences and make choices can't be explained by our physiological makeup and evolutionary history.

STEVEN PINKER is the Johnstone Family Professor in the Department of Psychology at Harvard University. He is the Author of *Language Learnability* and *Language Development; Learnability and Cognition; The Language Instinct; How the Mind Works; Words and Rules; The Blank State;* and *The Stuff of Thought.*

Why are empirical questions about how the mind works so weighted down with political and moral and emotional baggage? Why do people believe that there are dangerous implications to the idea that the mind is a product of the brain, that the brain is organized in part by the genome, and that the genome was shaped by natural selection? This idea has been met with demonstrations, denunciations, picketings, and comparisons to Nazism, from the right and the left. And these reactions affect both the day-to-day conduct of science and the public appreciation of the science. By exploring the political and moral colorings of discoveries about what makes us tick, we can have a more honest science and a less fearful intellectual milieu.

It's harder to find the truth if certain factual hypotheses are third rails—touch them and die. A clear example is research on parenting. Hundreds of studies have measured correlations between the practices of parents and the way their children turn out. For example, parents who talk a lot to their children have kids with better language skills, parents who spank have children who grow up to be violent, parents who are neither too authoritarian nor too lenient have children who are well adjusted, and so on. Most of the parenting-expert industry, and a lot of government policy, turn these correlations into advice to parents and blame the parents when children don't turn out as they would have liked. But correlation does not imply causation. Parents provide their children with genes as well as an environment, so the fact that talkative parents have kids with good language skills could simply mean that the same genes that make parents talkative make their children articulate. Until those studies are replicated with adopted children, who don't get their genes from the people who bring them up, we won't know whether the correlations reflect the effects of parenting, the effects of shared genes, or some mixture. But in most cases even the possibility that the correlations reflect shared genes is taboo. In developmental psychology, it's considered impolite even to mention it, let alone test it.

Most intellectuals today have a phobia about any explanation of the mind that invokes genetics. They're afraid of four things: First, there is a fear of inequality. The great appeal of the doctrine that the mind is a blank slate is the simple mathematical fact that zero equals zero. If we all start out blank, then no one can have more stuff written on his or her slate than anyone else. Whereas if we come into the world endowed with a rich set of mental faculties, these could work differently in people—better in some people than in others. The fear is that this would open the door to discrimination, oppression, eugenics, or even slavery and genocide. Of course, that is all a non sequitur. As many political writers have pointed out, commitment to political equality is not an empirical claim that people are clones. It's a moral claim that in certain spheres we judge people as individuals and don't take into account the statistical average of the groups they belong to. It's also a recognition that however much people might vary, they have certain things in common by virtue of their common human nature. No one likes to be humiliated or oppressed or enslaved or deprived. Political equality consists of recognizing, as the Declaration of Independence says, that people have certain inalienable rights—namely, life, liberty, and the pursuit of happiness. Recognizing those rights is not the same thing as believing that people are indistinguishable in every respect.

The second fear is the fear of imperfectability. If people are innately saddled with certain sins and flaws, like selfishness, prejudice, short-sightedness, and self-deception, then political reform would seem to be a waste of time. Why try to make the world a better place if people are rotten to the core and will just screw it up no matter what you do? People with sympathies for the romantic revolutionary politics of the 1960s and 1970s—which is where the initial opposition to sociobiology came from—have always been enraged by the claim that limitations on human nature might constrain our social arrangements. Again, this is a faulty argument. We know that there can be social improvement because we know that there *has* been social improvement—the end of slavery, torture, blood feuds, despotism, and the ownership of women in western democracies. Social change can take place even with a fixed human nature because the mind is a complex system of many parts. We may have motives that tempt us to do awful things; we also have motives that can counteract them. We can figure out ways to pit one human desire against another and thereby improve our condition, in the same way that we manipulate physical and biological laws (instead of denying that they exist) to improve our physical condition. We combat disease, we keep out the weather, we grow more crops—and we can jigger with our social arrangements as well.

A good example is the invention of democratic government. As Madison argued, by instituting checks and balances in a political system, one person's ambition counteracts another's. It's not that we have bred or socialized a new human being who's free of ambition. We have just developed a system in which these ambitions are kept under control.

Another reason that human nature doesn't rule out social progress is that many features of human nature have free parameters. This has long been recognized in the case of language: Some languages use the mirror image of the phrase-order patterns found in English but otherwise work by the same logic. Our moral sense may have a free parameter as well. People in all cultures are able to respect and sympathize with other people. The question is, With *which* other people? The default setting of our moral sense may be to sympathize only with members of our own clan or village. Over the course of history, a knob or a slider has been adjusted so that a larger and larger portion of humanity is admitted into the circle of people whose interests we consider as comparable to our own. From the village or clan, the moral circle has expanded to the tribe, the nation, and most recently to all of humanity, as in the Universal Declaration of Human Rights. This observation (originally from the philosopher Peter Singer) is an example of how we can enjoy social improvement and moral progress even if we are fitted with certain faculties, as long as those faculties can respond to inputs. In the case of the moral sense, the relevant inputs may be a cosmopolitan awareness of history and the narratives of other peoples, which allow us to project ourselves into the experiences of people who might otherwise be treated as obstacles or enemies.

The third fear is a fear of determinism: that we will no longer be able to hold people responsible for their behavior because they can blame it on their brain or their genes or their evolutionary history—the evolutionary-urge or killer-gene defense. The fear is misplaced for two reasons. One is that the silliest excuses for bad behavior have in fact invoked the environment rather than biology—such as the abuse excuse that got the Menendez brothers off the hook in their first trial, the "black rage" defense that was used in an attempt to exonerate the Long Island Rail Road gunman, the "pornography made me do it" defense that defense lawyers for rapists have tried. If there's a threat to responsibility, it doesn't come from biological determinism but from *any* kind of determinism, including childhood upbringing, mass media, and social conditioning. But none of these should be taken seriously. Even if there are parts of the brain that compel people to do things for various reasons, there are other parts of the brain that respond to the legal and social contingencies that we call "holding people responsible for their behavior." For example, if I rob a liquor store, I'll get thrown in jail, or if I cheat on my spouse, my friends and relatives and neighbors will think I'm a boorish cad and refuse to have anything to do with me. By holding people responsible for their actions, we are implementing contingencies that can affect parts of the brain and lead people to inhibit what they would otherwise do. There's no reason that we should give up that lever on people's behavior—namely, the inhibition systems of the brain—just because we're coming to understand more about the temptation systems.

The final fear is the fear of nihilism. If it can be shown that all of our motives and values are products of the physiology of the brain, which in turn was shaped by the forces of evolution, then (according to the fear) those motives and values would be shams, without objective reality. I wouldn't *really* be loving my child; all I would be doing is selfishly propagating my genes. Flowers and butterflies and works of art would not be truly beautiful; my brain just evolved to give me a pleasant sensation when a certain pattern of light hits my retina. The fear is that biology will debunk all we hold sacred. This fear is based on a confusion between two very different ways to explain behavior. What biologists call a *proximate* explanation refers to what is meaningful to me, given the brain I have. An *ultimate* explanation, in contrast, refers to the evolutionary processes that gave me a brain with the ability to have those thoughts and feelings. Yes, evolution (the ultimate explanation for our minds) is a short-sighted, selfish process, in which genes are selected for their ability to maximize the number of copies of themselves. But that doesn't mean that *we* are selfish and short-sighted—at least, not all the time. There's nothing that prevents the selfish, amoral process of natural selection from evolving a big-brained social organism with a complex moral sense. There's an old saying that people who appreciate legislation and sausages should not see them being made. The same is true of human values: Knowing how they were made can be misleading if you don't distinguish the process from the product. Selfish genes don't necessarily build a selfish organization.

So if people are afraid of human nature, what do they believe instead? I believe that there is a quasi-religious theory of human nature prevalent among pundits and intellectuals which includes both empirical assumptions about how the mind works and a set of values that people hang on those assumptions. The theory has three parts: I have already mentioned the doctrine of the Blank Slate—that we have no inherent talents or temperaments because the mind is shaped completely by the environment (parenting, culture, and society). The second is the myth of the Noble Savage—that evil motives are not inherent in people but spring from corrupting social institutions. The third is the Ghost in the Machine—that the most important part of us is somehow independent of our biology, so that our ability to have experiences and make choices can't be explained by our physiological makeup and evolutionary history.

These three ideas are increasingly being challenged by the sciences of the mind, brain, genes, and evolution. They are being upheld more for their moral and political uplift than for any empirical rationale. People think that these doctrines are preferable on moral grounds and that the alternative is a forbidden territory we should avoid at all costs.

But the Blank Slate has been undermined by a number of discoveries. One of them is a simple logical point: No matter how important learning and culture and socialization are, they don't happen by magic. There has to be innate circuitry that does the learning, that creates the culture, that acquires the culture, and that responds to socialization efforts. Once you try to specify what those learning mechanisms are, you're forced to posit a great deal of innate structure to the mind.

The Blank Slate has also been undermined by behavioral genetics, which has found that at least half the variation in personality and intelligence within a society comes from differences in the genes. The most dramatic example is that identical twins separated at birth have fantastic similarities in their talents and tastes. The Blank Slate has also been undermined by evolutionary psychology and anthropology. For example, despite the undeniable variation among cultures, we now know that there is a vast set of universal traits common to the world's 6,000 cultures. Also, evolutionary psychology has shown that many of our motives make no sense in terms of our day-to-day efforts to enhance our physical and psychological well-being but can be explained in terms of the mechanism of natural selection operating in the environment in which we evolved. A relatively uncontroversial example is our taste for sugar and fat, which were adaptive in an environment in which those nutrients were in short supply but don't do anyone any good in a modern environment, in which they are cheap and available anywhere. A more controversial example may be the universal thirst for revenge, which was one's only defense in a world in which one couldn't dial 911 to get the police to show up if one's interests were threatened. A belligerent stance was one's only deterrent against those whose interests were in conflict with one's own. A third is our taste for attractive marriage partners. As wise people have pointed out for millennia, physical appearance is not a good predictor of how happy or compatible a couple will be. The curve of your date's nose or the shape of her chin doesn't predict how well you're going to get along with her for the rest of your life. But evolutionary psychology has shown that the physical features of beauty are cues to health and fertility. Our fatal weakness for attractive partners can be explained in terms of our evolutionary history, not our personal calculations of well-being. The Blank Slate has also been undermined by brain science. The brain obviously has a great deal of what neuroscientists call plasticity—that's what allows us to learn. But the newest research is showing that many properties of the brain are genetically organized and don't depend on information coming in from the senses.

The doctrine of the Noble Savage has been undermined by a revolution in our understanding of non-state societies.

Many intellectuals believe that violence and war among hunter/gatherers are rare or ritualistic, and that a battle is called to a halt as soon as the first man falls. But studies that count the dead bodies have shown that the homicide rates among prehistoric peoples are orders of magnitude higher than those in modern societies—even taking into account the statistics from two world wars! We also have evidence that nasty traits such as psychopathy, violent tendencies, a lack of conscientiousness, and an antagonistic personality are to a large extent heritable. And there are mechanisms in the brain, probably shared across primates, that underlie violence. All these suggest that what we don't like about ourselves can't just be blamed on the institutions of a particular society.

And the Ghost in the Machine has been undermined by cognitive science and neuroscience. The foundation of cognitive science is the computational theory of mind—the idea that intelligence can be explained as a kind of information processing, and that motivation and emotion can be explained as cybernetic feedback systems. Feats and phenomena that were formerly thought to rely on mental stuff alone—such as beliefs, desires, intelligence, and goal-directed behavior—can be explained in physical terms. And neuroscience has most decisively exorcised the Ghost in the Machine by showing that our thoughts, feelings, urges, and consciousness depend entirely on the physiological activity of the brain.

Of the four new sciences of human nature—cognitive science, neuroscience, behavioral genetics, evolutionary psychology—evolutionary psychology has probably aroused the most controversy in the last decade, much of it needless.

There's a sense in which all psychology is evolutionary. When it comes to understanding a complex psychological faculty such as thirst or shape perception or memory, psychologists have always appealed to their evolutionary functions, and that's never been controversial. It's no coincidence that the effects of thirst are to keep the balance of water and electrolytes in the body within certain limits required for survival; without such a mechanism, organisms would plump up and split like a hot dog on a grill or shrivel up like a prune. Likewise, it can't be a coincidence that the brain compares the images from the two eyeballs and uses that information to compute depth. Without such an ability, we'd be more likely to bump into trees and fall off cliffs. The only explanation, other than creationism, is that those systems evolved because they allowed our ancestors to survive and reproduce better than the alternatives.

Evolutionary psychology is simply taking that mindset and applying it to more emotionally charged aspects of behavior, such as sexuality, violence, beauty, and family feelings. One reason that evolution is more controversial in these areas than it is in the study of thirst is that the implications of evolution are less intuitive in the case of emotions and social relations. You don't need to know much evolutionary biology to say that it's useful to have stereo vision or thirst. But when it comes to how organisms deal with one another, common sense is no substitute for serious evolutionary theory. We have no good intuitions about whether it's adaptive, in the narrow biologist's sense, to be monogamous or polygamous, to treat all your children equally or to play favorites, to be attracted to one kind of facial geometry or another. There you have to learn what the best evolutionary biology predicts. So evolutionary thinking in those fields is more surprising than in the rest of psychology.

Behavioral genetics has also challenged our intuitions. Here is a puzzle. We know that genes matter in the formation of personalities. Probably about half the variation in personality within a culture can be attributed to differences in genes. Upon hearing this, people often conclude that the other half must come from the way parents bring up their children: half heredity, half environment—a nice compromise, right? Wrong. The other 50 percent of the variation turns out not to be explained by which family you've been brought up in. Concretely, here's what behavioral geneticists have found. Everyone knows about the identical twins separated at birth who have remarkable similarities: They score similarly on personality tests, they have similar tastes in music, they have similar political opinions, and so on. But the other discovery, which is just as important though less well appreciated, is that twins separated at birth are no more different from each other than twins brought up together in the same house with the same parents, the same number of TV sets, the same number of books, the same number of guns, and so on. Growing up together doesn't make you and your twin more similar in intelligence or personality over the long run. A corroborating finding is that adopted siblings, who grow up in the same house but don't share genes, are not correlated in personality and intelligence at all; they are no more similar than two people plucked off the street at random. So it's not all in the genes, but what isn't in the genes isn't in the family environment, either. It can't be explained in terms of the overall personalities or the child-rearing practices of parents.

What are *non*-genetic determinants of personality and intelligence, given that they almost certainly are not the family environment? This puzzle was first noticed by behavioral geneticists such as David Rowe, Robert Plomin, and Sandra

Scarr, and has been the subject of recent books by Judith Rich Harris and Frank Sulloway. Many people, still groping for a way to put parents back into the picture, assume that differences among siblings must come from differences in the way parents treat each of their children. Forget it. The best studies have shown that when parents treat their kids differently, it's because the kids are different to begin with, just as anyone reacts differently to different people. Any parent of more than one child knows that children are little people, born with personalities.

Where Sulloway and Harris differ is that Sulloway argues that the unexplained variation comes from the way children differentiate themselves from their siblings in the family. They adopt strategies for competing for parental attention and resources outside the family and react to non-relatives by using the same strategies that worked for them inside the family. Harris argues that the missing variance comes from how children survive within peer groups—how they find a niche in their own society and develop strategies to prosper in it.

I think Sulloway has captured something about the dynamics among siblings within the family. But I'm not convinced that these strategies shape their personalities *outside* the family. What works with your little brother is not necessarily going to work with strangers and friends and colleagues. Most of the data supporting Sulloway come from studies in which siblings rate their siblings, or parents rate their children, or in which siblings rate themselves with respect to their siblings. The theory is not well supported by studies that look at the personality of people outside the home. Indeed, it's a major tenet of evolutionary psychology that one's relationships with kin are very different from one's relationships with non-relatives.

As for Harris, I'm persuaded by her argument that socialization takes place in the peer group rather than in the family. Most child psychologists won't go near that claim, but it survives one empirical test after another. To take a few examples: Children almost always end up with the accent of their peers, not their parents. Children of culturally inept immigrants do just fine if they can learn the ropes from native-born peers. Children who are thrown together without an adult language to learn will invent a language of their own. And many studies have shown that radical variations in parenting practice—whether you grow up in an Ozzie and Harriet family or a hippie commune, whether you have two parents of the same sex or one of each, whether you spent your hours in the family home or a day-care center, whether you are an only child or come from a large family, whether you were conceived the normal way or in a laboratory dish—leave no lasting marks on your personality as long as you are part of a normal peer group.

What Harris's theory has not explained to my satisfaction—at least, not yet—is the missing variation in personality per se. Personality and socialization aren't the same thing. Socialization is how you become a functioning person in a society—speak the language, win friends, hold a job, wear the accepted kinds of clothing. "Personality" is whether you're nice or nasty, bold or shy, conscientious or lackadaisical. Here's the problem. Let's go back to our touchstone: identical twins brought up together who share their genes and most of their environment but nonetheless are not identical in personality. They almost certainly will have grown up in the same peer groups, or at least the same kinds of peer groups, and their personalities and physical characteristics will tend to place them in the same niches within those peer groups. So peer groups by themselves can't explain the unexplained variation in personality. To be fair, Harris points out that which niche you fill in a peer group (the peacemaker, the loose cannon, the jester, the facilitator) might partly be determined by chance—which niche happens to be open when you find a circle of buddies to hang out with. There may be something to that, but it's a special case of what might be an enormous role for chance in the shaping of who we are. In addition to which niche was open in your peer group, other unpredictable events may affect each of us as we grow up. Did you get the top bunk bed or the bottom one? Did you get chased by a dog, or dropped on your head, or infected by a virus, or smiled on by a teacher?

And there are even more chance events in the wiring of the brain *in utero* and the first couple of years of life. We know that there isn't nearly enough information in the genome to specify the brain down to the last synapse, and that the brain isn't completely shaped by incoming sensory information, either. Based on studies of the development of simple organisms like fruit flies and roundworms, we know that much in development is a matter of chance. Among genetically homogeneous strains of roundworm brought up in the same monotonous laboratory conditions, one animal can live three times as long as another. Two fruit flies from inbred strains—in effect, clones—can be physically different: They can have different numbers of bristles under each wing, for example. If simple organisms like worms and flies can turn out differently for capricious reasons, then surely chance plays an even bigger role in the way our brains develop.

The idea that the human mind is a blank slate has had an enormous influence in many fields. One is architecture and urban planning. The twentieth century saw the rise of a

movement that has been called authoritarian high modernism, contemporaneous with the ascendance of the blank
slate. City planners believed that our tastes for green space,
for ornament, for people-watching, for cozy places for intimate social gatherings, were social constructions. They were
thought to be archaic historical artifacts that got in the way
of the orderly design of cities and should be ignored by planners designing optimal cities according to so-called scientific
principles. Le Corbusier was the clearest example. He and
other planners had a minimalist conception of human nature: A human being, they thought, needs so many cubic feet
of air per day, so many gallons of water, so many square feet
in which to sleep and work, a temperature within a certain
range, and so on. Houses became "machines for living," and
cities were designed around the most efficient ways to satisfy
this short list of needs—namely, freeways, huge rectangular
concrete housing projects, and open plazas. In extreme cases
this led to the wastelands of planned cities like Brasilia; in
milder cases it gave us the "urban renewal" projects in American cities, the dreary highrises in the Soviet Union, and
English council flats. Ornamentation, human scale, green
space, gardens, and comfortable social meeting places were
written out of cities because the planners had a theory of human nature that omitted human aesthetic and social needs.

Another example is the arts. In the twentieth century,
modernism and postmodernism took over, and their practitioners disdained beauty as bourgeois, saccharine, lightweight.
Art was deliberately made incomprehensible or ugly or
shocking—again, on the assumption that our predilections
for attractive faces, landscapes, colors, and so on were
reversible social constructions. This also led to an exaggera-
tion of the dynamic of social status that has always been part
of the arts. The elite arts used to be aligned with the economic

and political aristocracy. They involved displays of sumptuosity and the flaunting of rare and precious skills that only the idle rich could cultivate. But now that any schmo could afford a Mozart CD or go to a free museum, artists had to figure out new ways to differentiate themselves from the rabble. So art became baffling and uninterpretable—unless you had some acquaintance with arcane theory.

By their own admission, the humanities programs in universities, and institutions that promote new works of elite art, are in crisis. People are staying away in droves. I don't think it takes an Einstein to figure out why. By denying the human sense of visual beauty in painting and sculpture, of melody in music, of meter and rhyme in poetry, of plot and narrative and character in fiction, the elite arts wrote off the vast majority of their audience, the people who approach art in part for pleasure and edification rather than social one-upmanship. Today there are movements in the arts to reintroduce beauty and narrative and melody and other basic human pleasures. And those artists are considered radical.

Many artists and scholars have pointed out that ultimately art depends on human nature. The aesthetic and emotional reactions we have to works of art depend on how our brain is put together. Art works because it appeals to certain faculties of the mind. Music depends on details of the auditory system, painting and sculpture on the visual system. Poetry and literature depend on language. And the insights we hope to take away from great works of art depend on their ability to explore the eternal conflicts in the human condition, like those between men and women, self and society, parent and child, sibling and sibling, friend and friend. Some theoreticians of literature have suggested that we appreciate tragedy and great works of fiction because they explore the permutations and combinations of human conflict—and these are the very themes that fields like evolutionary psychology and behavioral genetics and social psychology try to illuminate. The sciences of the mind can reinforce the idea that there is an enduring human nature that great art can appeal to.

We may be seeing a coming together of the humanities and the science of human nature. They've long been separated, because of postmodernism and modernism. But now graduate students are grumbling in e-mails and conference hallways about being locked out of the job market unless they perpetuate postmodernist gobbledygook, and about how eager they are for new ideas from the sciences that will invigorate the humanities within universities. Connoisseurs and appreciators of art are getting sick of the umpteenth exhibit on the female body featuring mangled body parts, or ironic allusions to commercial culture that are supposed to shake people out of their bourgeois complacency but are really no more insightful than an ad parody in *Mad* magazine or on *Saturday Night Live*.

Intellectual life over the past century has been enormously affected by an understandable revulsion to Nazism, with its pseudoscientific theories of race and its equally nonsensical glorification of conflict as part of the evolutionary wisdom of nature. It was natural to reject anything that smacked of a genetic approach to human affairs. But historians of ideas have begun to fill in another side of the picture. The remarkable fact is that the two great ideologically driven genocides of the twentieth century came from theories of human nature that were diametrically opposed. The Marxists had no use for the concept of race, didn't believe in genes, and denied Darwin's theory of natural selection as the mechanism of evolutionary adaptation. It's not a biological approach to human nature that is uniquely sinister. There must be common threads to Nazism and totalitarian Marxism that cut across a belief in the importance of evolution or genetics. One common thread was a desire to reshape humanity. In the Marxists' case, it was through social engineering; in the Nazis' case, it was through eugenics. Neither was satisfied with human beings as we find them, with all their flaws and weaknesses. Rather than building a social order around enduring human traits, they thought they could reengineer human traits using scientific—in reality, pseudoscientific—principles.

In Martin Amis's recent book about Stalinism, he argues that intellectuals have not yet come to grips with the lessons of Marxist totalitarianism as they did many decades ago with Nazi totalitarianism. A number of historians and political philosophers have made the same point. This blind spot has distorted the intellectual landscape, including the implications and non-implications of genetics and evolution for understanding ourselves. Chekhov once said, "Man will become better when you show him what he is like." I can't put it any better than that.

GETTING HUMAN NATURE RIGHT

HELENA CRONIN

•

Certainly, human nature is fixed. It's universal and un-changing, common to every baby that's born, down through the history of our species. But human behavior, which is gen-erated by that nature, is endlessly variable and diverse. After all, fixed rules can give rise to an inexhaustible range of out-comes. Natural selection equipped us with the fixed rules— the rules that constitute our human nature. And it designed those rules to generate behavior that's sensitive to the envi-ronment. So the answer to genetic determinism is simple. If you want to change behavior, just change the environment. And to know which changes would be appropriate and effec-tive, you have to know those Darwinian rules. You need only to understand human nature, not to change it.

HELENA CRONIN is a codirector of the London School of Economics's Centre for Philosophy of Natural and Social Sciences, where she runs the wide-ranging and successful program called Darwin@LSE, which fosters research at the forefront of evolutionary theory. She is the author of *The Ant and the Peacock*.

The questions I'm asking myself at the moment are about the connections between two things. On the one hand, there's what science tells us about the evolved differences between women and men—what we know from modern Darwinian theory. And, on the other hand, there's the public perception of the science, which is largely negative and riddled with misunderstandings. Of course, when evolutionary theory gets applied to our own species, it always arouses opposition. But when it comes to sex differences—that sparks off hostilities and misconceptions all of its own.

It all stems from muddling science and politics. It's as if people believed that if you don't like what you think are the ideological implications of the science, then you're free to reject the science—and to cobble together your own version of it instead. Now, I know that sounds ridiculous; science doesn't have ideological implications, it simply tells you how the world is—not how it ought to be. So if a justification or a moral judgment or any such "ought" statement pops up as a conclusion from purely scientific premises, then obviously the thing to do is challenge the logic of the argument, not reject the premises. But unfortunately, people get so indignant about the conclusion that they end up rejecting the science rather than the fallacy.

The "implication" that seems to worry people most is so-called genetic determinism—the notion that if human nature was shaped by evolution, then it's fixed and we're simply stuck with it; there's nothing we can do about it. We can never change the world to be the way we want; we can never institute fairer societies—policy-making and politics are pointless.

Now, that's a complete misunderstanding. It doesn't distinguish between human nature—our evolved psychology—and the behavior that results from it. Certainly, human nature is fixed. It's universal and unchanging, common to every baby that's born, down through the history of our species. But human behavior, which is generated by that nature, is endlessly variable and diverse. After all, fixed rules can give rise to an inexhaustible range of outcomes. Natural selection equipped us with the fixed rules—the rules that constitute our human nature. And it designed those rules to generate behavior that's sensitive to the environment. So the answer to genetic determinism is simple. If you want to change behavior, just change the environment. And to know which changes would be appropriate and effective, you have to know those Darwinian rules. You need only to understand human nature, not to change it.

Margo Wilson and Martin Daly's classic work on homicide illustrates this clearly. Homicide rates vary enormously across different societies. In the 1970s and 1980s, when the rate in Chicago was 900 murders per million of the population per annum (for same-sex, non-kin killings), the rate in England and Wales was 30, and in Iceland there were hardly any murders at all. Now, there's no difference in the genes, no difference in human nature, in those places. That shows up very dramatically when you look at the patterns of the murders. Although the rates are vastly different, the patterns are exactly the same. If you shrink the axes of the Chicago graph of the age and sex of the murderers and lay it over the England/Wales graph, the curves are an exact fit. It's overwhelmingly young men killing young men—starting, peaking, and trailing off at exactly the same ages. What makes the difference in the rates is the different environments. And that's crucial for policy. We understand what it is about our evolved minds that leads to such different rates in different environments—the universal propensity of males to be highly competitive, which under extreme conditions can end up in homicide. And that tells us what conditions we need to create to lower the murder rates. Indeed, far from being genetic determinism, we can see why the Darwinian approach has even been called—with only a touch of irony— "an environmentalist discipline."

Genetic determinism fosters the notion that if genes are part of the causal process, then in order to change outcomes you've got to tweak the genes—you've got to alter that one particular cause. That's a very odd idea. There's no reason why you can't intervene at any part of the causal process, no reason that genes should take precedence. As we've seen with murder rates, when you're dealing with the universals of human nature, the environment is the obvious place to intervene. But that can also be true even when you're dealing with genetic differences between people. There are genetic differences, for example, in the propensity to develop adult diabetes. In an environment in which people eat traditional food—low calorie-density, high fiber, low fat, low sugar— nobody develops this kind of diabetes. But expose these populations to a modern diet and the people with the greater hereditary disposition show up immediately. Similarly, there could be genetic differences in men's disposition to compete. But in appropriate environments—more Iceland than Chicago—those differences would barely show up in the homicide statistics.

There are lots of other notions packed into genetic determinism—to do with free will and responsibility, control over your life, and so on. But I've yet to discover a single in terpretation of genetic determinism that carries any of the implications people seem to worry so much about. On the contrary, it turns out that whatever applies to genes applies equally to environments. So if people fear genetic determinism, they should be worrying equally about environmental determinism.

This kind of thinking applied to sex differences has led to deep hostility to the very idea of evolved differences between women and men. Feminists in particular have led this opposition. Of course, "feminism" covers a multitude of views. There's often not much in common between the unreconstructed Marxists of the British Left, the "postmodern" jargon-generators, and the CEO who's flicking shards of glass ceiling from her padded shoulders. But one thing on which most schools of feminism agree is that they're anti-Darwinian. Even the so-called "difference" feminists, who "celebrate" "us" versus "them," prefer to invent differences rather than defer to science. I find it all very dismaying—and, as a Darwinian and a feminist, doubly dismaying.

I think this retrenchment stems from a vague belief that you can't have fairness without sameness. I say "vague" because once you say it, you can see it's obviously false. But most strands of feminism have somehow got themselves committed to the view that if men and women are in any

ways fundamentally different, it will undermine the quest for
a fair and egalitarian society. What originally inspired feminism was the idea that women shouldn't be discriminated
against *qua* women—where it was irrelevant that they were
women: being barred from attending universities or owning
property or having the vote not because they were incapable
but because they were women. But that original inspiration
ends up seriously distorted when you deny evolved sex differences. Things have got to the point where there's expected to be some kind of fifty/fifty representation of men
and women everywhere—universities, workplace, politics,
sport, child care. So if women are not equally represented,
it's put down to sexism alone. Well, whether or not sexism is
operating, evolved sex differences certainly will be—differences in dispositions, skills, values, interests, and ambitions.
Women are very likely to make systematically different
choices from men. And it's those different preferences, not
blanket fifty/fifty distributions, that we should expect fair
policies to reflect.

Evolved sex differences are largely to do with averages.
So they don't cleave our species neatly in two. This is often
seized on as anti-Darwinian ammunition. I'm sure you've
heard the argument: "But the differences *within* the sexes are
greater than the differences *between* them." The implication
is that there's so much overlap in the distributions that the
Darwinian interest in differences is misleading.

But is that right? Whenever I try to think through exactly
what is being claimed, the argument tends to fall apart. For a
start, how important the difference is depends on why you're
interested in it, what your aim is. If your aim is to get rich,
don't try selling pornography to women or romantic novels
to men; don't try selling "Kill! Kill!" computer games to girls
or "people" games to boys. And, anyway, you can't simply

generalize about how large the overlap is; it depends on the characteristic. There'll be almost no overlap if you match boys against girls in throwing (the boys will win almost every time) or in fluency of speech (up to nine out of ten men will do worse than women). Then there's the fact that even if the mean differences are small, there can be huge differences at the extremes. Men are on average only a few inches taller than women, but all the very tallest people are male. So men might end up ahead just for that statistical reason alone.

There's also a curious fact—one uncovered by evolutionary biology—about the shapes of the distribution curves for most male/female differences. Darwin remarked on it, and on the fact that it holds robustly across other species, too. It's that males are far more variable than females: They are over-represented both at the top of the heap and the bottom of the barrel. For some characteristics, people might not care. But what about this implication? Fewer women are likely to be dunces, but also fewer will be geniuses. When I mentioned this in a seminar in the United States, I was sharply corrected by a group of feminists: "There's no such thing as genius!" I later discovered that this had become a fairly standard "feminist studies" line. I couldn't help wondering whether genius had been airbrushed out because there weren't many women in the picture. Darwinian theory also suggests that it's important to look at differences in disposition and interests. Will the top piano student become the international star? Being competitive, risk-taking, status-conscious, dedicated, single-minded, persevering—it can make all the difference to success. And these are qualities that men, on average, are far more likely to possess, often in alarming abundance.

Although "differences within and between" is a popular argument with feminists, it doesn't always fit happily with other feminist arguments. If there are wide "differences within," then women aren't very homogenous—there's a wide spread of abilities and dispositions—and some proportion of women will be in the male end of the distribution. That might be for any characteristic, from hormone levels to 3-D mental rotation (being able to imagine rotating objects in space, a notoriously male trick). But how does this mesh with the idea that women who are high achievers in traditionally male pursuits—engineering, mountaineering, or whatever—are "role models" for other women? The idea is that these women are just like the others and it's only male prejudice and self-doubt that hold the other women back. But maybe these women are the extremes of those "differences within" that feminists themselves emphasize, and so they're not just like the next woman? But then how can feminists confidently claim that it's only prejudice and self-doubt preventing any woman from achieving the same?

Worse, how can anyone confidently point to these women—as anti-Darwinians often do—as evidence against evolved sex differences? Far from undermining an evolutionary analysis, these women are probably exceptions that prove the Darwinian rule. So, for example, with 3-D mental rotation, women exposed in the womb to high levels of androgen perform far better than other women—indeed, almost as well as men. And with dispositions, too: Women in traditionally male professions respond to challenges with a characteristically "male" high adrenaline charge, and their job choice seems to follow their disposition rather than (as I wrongly guessed when I first heard this) their disposition being shaped by the job.

A final example: "Within and between" is used routinely to remind people like me that sex differences are only statistical generalizations and don't hold true for all individuals—which is, of course, right. But isn't the glass ceiling "only" a statistical generalization? There's an overlap in men's and women's jobs, particularly in middle management, and women aren't uniformly absent from high-flying posts. But is that a reason for dismissing the glass ceiling as unimportant? Statistical generalizations are exactly what many feminist issues are all about.

I think that the statistical distribution of male/female differences is a really interesting issue, with important implications for policy. It's one of those areas that's just waiting for the marriage of the evolutionary approach (which deals with universals) and behavioral genetics (which deals with individual differences). I'm really keen to see research on this. It seems to me to be something that Darwinism, feminism, and policy-makers most definitely need to deal with. Meanwhile, "within and between" gets us nowhere; it is useless—even downright misleading—as a guide to making decisions.

The mention of policy tends to provoke the question "But why drag in Darwin?" The question should, however, be the other way round. How could responsible social policy *not* be informed by an evolutionary understanding of sex differences? All policy-making should incorporate an understanding of human nature, and that means both female and male nature. Remember that if policy-makers want to change behavior, they have to change the environment appropriately. And what's appropriate can be very different for women and for men. Darwinian theory is crucial for pointing us to those differences.

I heard an American comedian the other day taking a swipe at "creeping neo-Darwinism." "I don't believe in the criminal gene," he said, "but if there was one, I think they'd find it right next to the out-of-work one." All very politically correct. But dead wrong on the differential impact of unemployment on men and women. For a woman, unemployment means loss of a job; for a man, it means loss of status. And this difference combines with other sex differences to take women and men down very different pathways once the workplace door closes on them. So, for example: A low-status man is a low-status mate; he'll have more difficulty finding a partner. And more difficulty keeping one; couples in which the wife earns more than the husband are more likely to divorce. He'll also be more at risk of "his" children not being his; misattributed paternity is as low as 1 percent among very high-status American males but up to 30 per-cent among unemployed, deprived, inner-city males. And then there's the risk of domestic violence; it stems from male sexual jealousy, and low status is a potent factor for moving the psychological machinery of jealousy into high gear. What's more, as in many other species, moving down in sta-tus has a grim impact on male (but not female) health and longevity. And, again as in other species, when the future looks inauspicious, males (but not females) are more likely to take risks. If "criminal genes" turn up next to "unemployment genes" in men, it's because a distinctive male psychology is forging the links. Anyone who really cares about unemploy-ment and its appalling social ramifications shouldn't be sniping at evolutionary theory; they should be embracing it. It's absolutely indispensable for getting a handle on the rel-evant causal connections.

Sex-blind social policy isn't impartial, isn't more fair—it's less so. Why, for example, assume that girls and boys learn in the same way? If you look at, say, mathematics, the academic area in which sex differences are most extreme, the boys' ad-vantage probably rests on their innate superiority in mechan-ical and three-dimensional thinking. There's some evidence that girls improve considerably if they're taught in ways that circumvent this. That's the kind of consideration that a fair education policy should be concerned with. And the same goes for the law, for the workplace, for economic planning— for whatever field social policy is being devised.

Our social policies need to cope with a world that is rapidly changing, and those changes include the relations between the sexes. There's the increase in male unemployment. There's women finally having the resources to go it alone as parents. And women finding that as their own status rises, the pool of potential partners shrinks. There are increasing inequalities, consigning substantial proportions of men to permanently low status. And there's growing acceptance that legal systems should not treat women as the chattel of men. How will our evolved psychology, our Stone Age minds, react to these changes? What will be significant for men and for women? So can Darwinian theory make a contribution to social policy? How could it not?

I'm aware that what I say is considered controversial, but it shouldn't be. I'm just doing standard science—and making the modest plea that policy should be based on knowledge. Indeed, the verdict should be the other way round. It's people who are prepared to talk about policy and society without knowing the first thing about human nature who should be considered controversial. But, sadly, science is widely undervalued. I think that one reason is the familiar scourge of relativism (particularly in its recent incarnations—postmodernism and its stablemates). Apart from the sciences, which have built-in immunity, it has taken a frightening hold on academia—on people who are influential and who are teaching future generations of influential people. The resulting attitudes toward science are deplorable—the view that there are no universal standards by which to judge truth or falsity or even logical validity, that science doesn't make progress, that there's nothing distinctive about scientific knowledge, and so on. One of the reasons why so much logic-free, fact-free, statistics-free criticism of Darwinism has been able to find an audience is this attitude that "Science is just another view, so I'm free to adopt my view, any view."

To make matters worse, this attitude tends to be viewed as liberal and open-minded. Science then comes to be seen, by contrast, as authoritarian and triumphalist. But science is characterized above all by its critical method. When scientists disagree, there are objective ways of deciding between them. Theories must be testable and then must pass the tests. On a day-to-day basis, matters won't always be clear-cut; science is not an instant process. Neither, of course, is it infallible. But it's by far the best we've got and it's done a breathtakingly impressive job so far. Once people understand what the scientific method is about and why it is so powerful, they will begin to appreciate that there really is a vast distinction between science and nonscience.

Mind you, the power of evolutionary theory isn't properly appreciated even within science. It's now a century and a half after the publication of the *Origin* and still Darwinian theory hasn't penetrated into many areas of biology. Even among biologists who do take an adaptationist approach, all too many drop it rather hastily when it comes to our own species—particularly when it comes to our psychology and our behavior, and most of all when it comes to sex differences. I'm often reminded of the anti-Darwinian attitudes of the nineteenth and early twentieth century—the period that has been called "the eclipse of Darwinism." Biology was rife with vulgar empiricism, dismissing adaptationist explanations on the grounds that they were teleological, went beyond the evidence, and so weren't genuine science.

So the problem is not only with the public's perception of Darwinism and sex differences. Many a scientist has also yet to be persuaded. But while the earlier rejection of Darwinism was rather tragic, this one's looking increasingly like farce. It's clear which way the history of science is going from here.

NATURAL-BORN CYBORGS?

ANDY CLARK

Our brains are (by nature) unusually plastic; their biologically proper functioning has always involved the recruitment and exploitation of nonbiological props and scaffolds. More so than any other creature on the planet, we humans emerge as natural-born cyborgs, factory-tweaked and primed so as to be ready to grow into extended cognitive and computational architectures—ones whose systemic boundaries far exceed those of skin and skull.

ANDY CLARK is professor of philosophy and director of the Cognitive Science Program at Indiana University. He was previously professor of philosophy at Sussex University, UK, and director of the Philosophy/Neuroscience/Psychology Program at Washington University in St. Louis. He is the author of *Microcognition*; *Associative Engines*; *Being There*; *Mindware*; and *Natural-Born Cyborgs*.

My body is an electronic virgin. I incorporate no silicon chips, no retinal or cochlear implants, no pacemaker. I don't even wear glasses. But I am slowly becoming more and more a cyborg. So are you. Pretty soon, and still without the need for wires, surgery, or bodily alterations, we shall be kin to the Terminator, to Eve 8, to Cable. . . . Just fill in your favorite fictional cyborg. Perhaps we already are. For we shall be cyborgs not in the superficial sense of combining flesh and wires but in the more profound sense of being human/technology symbionts—thinking and reasoning systems whose minds and selves are spread across biological brain and nonbiological circuitry.

This may sound like futuristic mumbo-jumbo, and I happily confess that I wrote the preceding paragraph with an eye to catching your attention, even if only by the somewhat dangerous route of courting your immediate disapproval! But I do believe that it is the plain and literal truth. I believe that it is above all a scientific truth, a reflection of some deep and important facts about (a whiff of paradox here?) our special and distinctively human nature. And certainly I don't think this tendency toward cognitive hybridization is a modern development; rather, it is an aspect of our humanity which is as basic and ancient as the use of speech and has been extending its territory ever since.

We see some of the "cognitive fossil trail" of the cyborg trait in the historical procession of potent cognitive technologies that begins with speech and counting, morphs first into written text and numerals, then into early printing (without movable typefaces), on to the revolutions of movable typefaces and the printing press, and most recently to the digital encodings that bring text, sound, and image into a uniform and widely transmissible format. Such technologies, once up and running in the various appliances and institutions that surround us, do far more than merely allow for the external storage and transmission of ideas. They constitute a cascade of mindware upgrades—cognitive upheavals in which the effective architecture of the human mind is altered and transformed.

What's more, the use, reach, and transformative powers of these cognitive technologies are escalating. New waves of user-sensitive technology may soon bring this ancient process to a climax, as our minds and identities become ever more deeply enmeshed in a nonbiological matrix of machines, tools, props, codes, and semi-intelligent daily objects.

We humans have indeed always been adept at dovetailing our minds and skills to the shape of our current tools and aids. But when those tools and aids start dovetailing back—when our technologies actively, automatically, and

continually tailor themselves to us, just as we do to them—
the line between tool and user becomes flimsy indeed. Such
technologies will be less like tools and more like part of the
mental apparatus of the person. They will remain tools only
in the thin and ultimately paradoxical sense in which my own
unconsciously operating neural structures (my hippocampus,
my posterior parietal cortex) are tools. I do not really "use"
my brain; rather, the operation of the brain is part of what
makes me who and what I am. So too with these new waves
of sensitive, interactive technologies. As our worlds become
smarter and get to know us better and better, it becomes
harder and harder to say where the world stops and the person begins.

What are these technologies? They are many and various. They include potent, portable machinery linking the
user to an increasingly responsive World Wide Web. But
they include also, and perhaps ultimately more important,
the gradual smartening-up and interconnection of the many
everyday objects populating our homes and offices.

My immediate goal, however, is not to talk about new
technology but to talk about us—about our sense of self and
the nature of the human mind. The point is not to guess at
what we might soon become, but to better appreciate what
we already are: creatures whose minds are special precisely
because they are tailor made to mix and match neural, bodily, and technological ploys.

Cognitive technologies are best understood as deep and
integral parts of the problem-solving systems that constitute
human intelligence. They are best seen as parts of the computational apparatus that constitutes our minds. If we do not
always see this, or if the idea seems outlandish or absurd, that
is because we are in the grip of a simple prejudice: the prejudice that whatever matters about the mind must depend

solely on what goes on inside the biological skin-bag, inside the ancient fortress of skin and skull. But this fortress has been built to be breached. It is a structure whose virtue lies in part in its ability to delicately gear its activities to collaborate with external, nonbiological sources of order so as (originally) to better solve the problems of survival and reproduction.

Consider a brief but representative example—the familiar process of writing an article for a newspaper, an academic paper, a chapter in a book. Confronted at last with the shiny finished product, we may find ourselves congratulating our brain on its good work. But this is misleading. It is misleading not simply because (as usual) most of the ideas were not our own anyway, but because the structure, form, and flow of the final product often depend heavily on the complex ways in which the brain cooperates with and depends on various special features of the media and technologies with which it continually interacts. We tend to think of our biological brains as the point source of the whole final content. But if we look a little more closely, what we may often find is that the biological brain participated in some potent and iterated loops through the cognitive technological environment.

We began perhaps by looking over some old notes, then turned to some original sources. As we read, our brain generated a few fragmentary, on-the-spot responses, which were duly stored as marks on the page or in the margins. The cycle repeats, pausing to loop back to the original plans and sketches, amending them in the same fragmentary, on-the-spot fashion. This whole process of critiquing, rearranging, streamlining, and linking is deeply informed by specific properties of the external media, which allow the sequence of simple reactions to become organized and grow into something like an argument. The brain's role is crucial and special. But it is not the whole story.

In fact, the true power and beauty of the brain's role is that it acts as a mediating factor in a variety of complex and iterated processes which continually loop between brain, body, and technological environment. And it is this larger system which solves the problem. We thus confront the cognitive equivalent of Richard Dawkins's vision of the extended phenotype. The intelligent process *is* just the spatially and temporally extended one which zigzags between brain, body, and world.

One useful way to understand the cognitive role of many of our self-created cognitive technologies is as affording complementary operations to those that come most naturally to biological brains. Consider the connectionist image of biological brains as pattern-completing engines. Such devices are adept at linking patterns of current sensory input with associated information: You hear the first bars of the song and recall the rest; you see the rat's tail and conjure the image of the rat. Computational engines of that broad class prove extremely good at tasks such as sensorimotor coordination, face recognition, voice recognition, and so on. But they are not well suited to deductive logic, planning, and the typical tasks of sequential reason. They are, roughly speaking, good at Frisbee, bad at logic—a cognitive profile at once familiar and alien. Familiar because human intelligence clearly has something of that flavor, yet alien because we repeatedly transcend these limits, planning family vacations, running economies, solving complex sequential problems, and so forth.

A powerful hypothesis—which I first encountered in work by the cognitive scientists David Rumelhart, Paul Smolensky, John McClelland, and Geoffrey Hinton—is that we transcend these limits in large part by combining the internal operation of a connectionist pattern-completing device with a variety of external operations and tools that serve to reduce various complex sequential problems to an ordered set of simpler pattern-completing operations, of the kind our brains are most comfortable with. Thus, to borrow their illustration, we may tackle the problem of long multiplication—say, 667 x 999—by using pen, paper, and numerical symbols. We then engage in a process of external symbol manipulations and storage, so as to reduce the complex problem to a sequence of simple pattern-completing steps we already command, first multiplying 9 by 7 and storing the result on paper, then 9 by 6, and so on.

The cognitive anthropologist Edwin Hutchins, in his book *Cognition in the Wild*, depicts the general role of cognitive technologies in similar terms, suggesting that such tools "permit the [users] to do the tasks that need to be done while doing the kinds of things people are good at: recognizing patterns, modeling simple dynamics of the world, and manipulating objects in the environment." This description nicely captures what is best about good examples of cognitive technology: recent word-processing packages, Web browsers, mouse and icon systems, and the like. (It also suggests, of course, what is wrong with many of our first attempts at creating such tools; the skills needed to use those environments—early VCRs, word processors, et cetera—were precisely those that biological brains find hardest to support, such as the recall and execution of long, essentially arbitrary sequences of operations.)

The conjecture, then, is that one large jump or discontinuity in human cognitive evolution involves the distinctive way in which human brains repeatedly create and exploit various species of cognitive technology so as to expand and reshape the space of human reason. We, more than any other creature on the planet, deploy nonbiological elements (instruments, media, notations) to complement (but not, typically, to replicate) our basic biological modes of processing, thereby creating extended cognitive systems whose computational and problem-solving profiles are quite different from those of the naked brain. Human brains maintain an intricate cognitive dance with an ecologically novel and immensely empowering environment: the world of symbols, media, formalisms, texts, speech, instruments, and culture. The computational circuitry of human cognition thus flows both within and beyond the head.

Such a point is not new and has been well made by a variety of theorists working in many different traditions. I believe, however, that the idea of human cognition as subsisting in a hybrid extended architecture—one which includes aspects of the brain and of the cognitive technological envelope in which our brains develop and operate—remains vastly underappreciated. We simply cannot hope to understand what is special and distinctively powerful about human thought and reason by merely paying lip service to the importance of this web of surrounding technologies. We need to work towards a much more detailed understanding of how our brains actively dovetail their problem-solving activities with a variety of nonbiological resources, and how the larger systems thus created operate, change, interact, and evolve. In addition, it may soon be important (morally, socially, and politically) to publicly loosen the bonds between the very ideas of minds and persons and the image of the bounds, properties, locations, and limitations of the basic biological organism.

A proper question to press is this: Since no other species on the planet builds as varied, complex, and open-ended designer environments as we do (the claim, after all, is that this is why we are special), what allowed this process to get off the ground in our species in such a spectacular way? And isn't that, whatever it is, what really matters? Otherwise put, even if it's the designer environments that make us so intelligent, isn't it some deep biological difference that lets us build/discover/use them in the first place?

This is a serious, important, and largely unresolved question. Clearly there must be some (perhaps quite small) biological difference that lets us get our collective foot in the designer-environment door. What can it be? One possible story locates the difference in a biological innovation for widespread cortical plasticity combined with the extended period of protected learning called childhood. Thus, neural constructivists such as Steve Quartz and Terry Sejnowski depict neural (especially cortical) growth as experience-dependent and as involving the actual construction of new neural circuitry (synapses, axons, dendrites) rather than just the fine-tuning of circuitry whose basic shape and form are already determined. One upshot is that the learning device itself changes as a result of interactions between organism and environment. Learning does not just alter the knowledge base for a fixed computational engine; it alters the internal computational architecture itself. The linguistic and technological environments in which human brains grow and develop are thus poised to function as the anchor points around which such flexible neural resources adapt and fit.

Perhaps, then, it is a mistake to posit a biologically fixed "human nature" with a simple wraparound of tools and culture, for the tools and culture are as much determiners of our nature as products of it. Our brains are (by nature) unusually plastic; their biologically proper functioning has always involved the recruitment and exploitation of nonbiological props and scaffolds. More so than any other creature on the planet, we humans emerge as *natural-born cyborgs*, factory-tweaked and primed so as to be ready to grow into extended cognitive and computational architectures—ones whose systemic boundaries far exceed those of skin and skull.

All this adds interesting complexity to those evolutionary psychological accounts that emphasize our ancestral environments. For we must now take into account an exceptionally plastic evolutionary overlay which yields a constantly moving target, an extended cognitive architecture whose constancy lies mainly in its continual openness to change. Even granting that the biological innovations which got this ball rolling may have consisted only in some small tweaks to an ancestral repertoire, the upshot of this subtle alteration is a sudden and enormous leap in cognitive-architectural space. Our cognitive machinery is now intrinsically geared to transformation, technology-based expansion, and a snowballing and self-perpetuating process of computational and representational growth. The machinery of contemporary human reason is rooted in a biologically incremental progression, while simultaneously existing on the far side of a precipitous cliff in cognitive-architectural space.

In sum, the project of understanding human thought and reason is easily and frequently misconstrued. It is misconstrued as the project of understanding what is special about the human brain. No doubt there *is* something special about our brains. But understanding our peculiar profiles as reasoners, thinkers, and knowers of our worlds requires an even broader perspective: one that targets multiple brains and bodies operating in specially constructed environments replete with artifacts, external symbols, and all the variegated scaffoldings of science, art, and culture.

Understanding what is distinctive about human reason

involves understanding the complementary contributions of both biology and (broadly speaking) technology, as well as the dense reciprocal patterns of causal and co-evolutionary influence that run between them. We cannot see ourselves aright until we see ourselves as nature's very own cyborgs— cognitive hybrids who repeatedly occupy regions of design space radically different from those of our biological forebears. The hard task, of course, is now to transform all this from mere impressionistic sketch into a balanced scientific account of the extended mind.

ANIMAL MINDS

Marc D. Hauser

In my own work, we've begun looking at the kinds of computations that animals and human infants are capable of when they interact with the physical and social world. We want to understand how such capacities evolved and how they constrain thought.

MARC D. HAUSER is a cognitive neuroscientist at Harvard University, where he is a Harvard College Professor, a professor in the Department of Psychology and the Program in Neurosciences, and director of the Mind, Brain, and Behavior Program. He is author of *The Evolution of Communication; Wild Minds;* and *Moral Minds.*

———

Some of the problems we've been dealing with in the neurosciences and cognitive sciences concern the initial state of the organism. What do animals, including humans, come into this world equipped with? What mental tools do they have to cope with the problems of the physical and social world? There's something of an illusion in the neurosciences that we have begun to really understand how the brain works. Noam Chomsky, in a recent talk entitled "Language and the Brain," warned neuroscientists about how little we know, especially when it comes to understanding how the brain handles language.

Here's the idea Chomsky played with, which I think is correct, and forms an essential part of the approach I take in my research. In looking at any cognitive system, you want to ask three questions. First, what constitutes knowledge in a particular domain, such as language or music or morality? Second, how is such knowledge acquired? Third, how is such knowledge used in the world? Let's take a very simple system that's good at a kind of computation based on particular knowledge of the world: the honeybee. This tiny insect—tiny brain, simple nervous system—is capable of transmitting information to its colony about where it's been and what it's eaten, and that information is sufficiently precise that the colony members can go out and find the food. We know that kind of information is encoded in the signal because of findings from a robotic honeybee, programmed to dance in a certain way and replicate real honeybee behavior; you can plop this robot in the middle of a colony, set it dancing in honeybee contra-dancing style, and the hive members will take the information and zip off to the designated location. But when you step back and ask, "What do we know about how the *brain* of the honeybee represents this kind of information?" the answer is, "We know almost nothing." Our understanding of the way in which a bee's brain represents its dance—that is, its language—is poor. And yet we are looking at a relatively simple nervous system, especially compared with the human nervous system. This conclusion in no way undermines the progress that honeybee researchers have made in documenting what honeybees know about the world, how they acquire it, and how they deploy it. What's missing, or at least poorly understood, is how the honeybee brain represents what it knows, and how the brain acquires and deploys this information.

Chomsky's main point is that what we know about how the human brain represents language is, at some level, trivial. Neuroscientists have made a lot of progress, in that we know what areas of the brain, when damaged, will wipe out certain aspects of the faculty of language; for example, damage to a particular part of the brain results in the loss of representations for consonants, while other damage results in the loss of representations for vowels. But we know relatively little

about how the circuitry of the brain represents the consonants and vowels. The chasm between the current neurosci-
entific understanding of the brain and understanding
representations like language is very wide.

A related point concerns how the internal computations
and the mechanisms underlying acquisition of knowledge
evolved. Consider language again. We can ask whether other
animals share that faculty with us. If not, is it because they lack
the internal computations or because of constraints that lie
outside the faculty of language proper, such as insufficient
memory or capacity to imitate? In the primates, the frontal
lobes of the brain, which play a role in short-term storage of
representations, have undergone an enormous change over
time. Therefore, our closest living relatives, the apes, probably
don't have the neural structures that would allow them to do
the kind of computations you need to do language processing,
including keeping a long string of utterances in mind to
process meaning. In my own work, we've begun looking at the
kinds of computations that animals and human infants are capable of when they interact with the physical and social world.
We want to understand how such capacities evolved and how
they constrain thought.

Whenever nature has created systems that seem open-
ended and generative, they've used a discrete set of recom-
binable elements. The question you can ask in biology is,
"What kinds of systems are capable of those kinds of computational processes?" Many organisms seem to be capable of
simple statistical computations, such as conditional probabilities that focus on local dependencies: *If A, then B.* Lots of
animals seem capable of that. But when you step up to the
next level in the computational hierarchy—one that requires
recursion—you find great limitations among both animals
and human infants. For example, an animal that can do *if A,*

then B would have great difficulty doing *if A to the n, then B to the n.* We now begin to have a loop, a rule that refers to itself and generates a relatively unlimited range of expressions. If animals lack this capacity, which they seem to, then we have identified an evolutionary constraint. Humans have evolved the capacity for recursion, a computation that greatly liberated us, allowing us to do mathematics as well as language. And this system of taking discrete elements and recombining them is what gives genetics and chemistry their open-ended structure. Given this pattern, the interesting questions are: What were the selective pressures that led to the evolution of a recursive system? Why is it that humans seem to be the only organisms on the planet, the only natural system, that have this capacity? What were the pressures that created it?

With regard to artificial intelligence, what kinds of pressures on an artificial system would get it to that end point? An interesting problem for natural biological systems and artificial systems is whether the two can meet. What kinds of pressures lead to a capacity for recursion? Comparative biology doesn't provide any helpful hints at present, because we simply have two end points—humans that do it and other organisms that appear not to. This evolutionary transition is still opaque.

The big questions on my mind are those we have no answers for: questions like "Why is *Homo sapiens* the only species that sheds tears when it cries?" Emotions that provoke tears are common to both humans and animals, yet we're the only species that generates a physical output of those emotions. If you look at weeping from an evolutionary perspective, which has really not been done, you begin to get some of the answers. Unlike all the other emotional expressions, weeping leaves a long-term physical trace. It blurs your vision; therefore it's costly. It's also very difficult to fake. What this suggests is an idea that the evolutionary biologist Amotz Zahavi proposed many years ago: Signals that are costly to produce are honest signals; you can look at a signal and infer its honesty based on the cost of expression. Weeping is potentially one of those; it's even necessary for actors to actually experience the feeling before they can generate the expression, and even then it's hard to do naturally. We know that animals experience sadness; whether they experience joy is hard to say, but they certainly have the emotions that would accompany crying with tears, even if they don't have that connection in the brain. It's not that they don't shed tears, because they do if the eye is physically irritated; it's that they lack some neural connection between the psychological state underlying the emotion and the connection to the system that creates tears. To say they lack the connection in the brain is an answer at a single level of analysis, the level of mechanism: What brain mechanisms support weeping? It's more interesting to take the evolutionary approach and ask why we cry with tears and other animals do not. And the answer is that weeping is an expression that conveys honesty.

For the past few years, I have been using the theoretical tools of evolutionary biology to ask questions about the design of animal minds. The notion of the environment for evolutionary adaptedness being pegged to the hunter/gatherer period in the Plio-Pleistocene may be true for some aspects of the human mind but it's probably wrong for many other aspects. How do organisms navigate through space; how do they recognize what is an object; how do they enumerate objects in their environment? Those aspects are probably shared across a wide variety of animals. Instead of deciding that the human mind evolved and was shaped during the Plio-Pleistocene, it's more appropriate to ask what happened in the Plio-Pleistocene that would have created a particular signature to the human mind which doesn't exist in other animals.

I've been looking at various domains of knowledge and asking what selective pressures shaped the way different organisms think. I'm trying to get away from the common approach to thinking about humans, human evolution, and animal cognition—that humans are unique and that's the end of the story. *All* animals are unique, and the really interesting question is how their minds have been designed by the particular social and ecological problems the environment throws at them. For example, instead of stating that humans are unique, we ask: What pressures did humans confront that no other animal confronted and which created selection for the evolution of language? Why can other organisms make do with the kinds of communication systems they have? Why did we evolve color vision? Why did other organisms not evolve color vision? Why are certain animals able to navigate in space with a simple mechanism like dead reckoning, while other animals need other kinds of machinery in order to get by in space? Why might we be the only, or one of the few, animals that have the ability to make inferences about what other people believe and desire?

This approach to the study of animals and humans brings the two disciplines together for the first time, armed with new comparative methods. We're now entering a period in the study of animal minds when we can use techniques that in part have been developed in the study of humans, especially human infants; conversely, methods developed on animals are being used by cognitive scientists who study humans. Here's one example: Researchers studying cognitive development, such as Susan Carey, Elizabeth Spelke, and Renee Baillargeon, have implemented a novel technique for asking human infants—who, of course, lack a functional linguistic system—how they think about the world. The technique is simple, really just a bit of magic. The idea is that when we watch magic shows, such as those of the great Houdini or David Copperfield, we become engaged because the magician is creating violations before our very eyes; at least they are violations based on the expectations we generate about the physical world. For instance, human bodies are not capable of being cut in half and put back together again. If the logic of a magic show or the special effects in a movie grab our attention, it's precisely because our expectations have been violated. We can ask what expectations infants or nonhuman animals bring into the world about how things should work, and the extent to which particular kinds of experience alter their expectations. If they, too, have specific expectations, then we should be able to create a magic show and grab their attention. They should show more interest in a magic show than in a comparable demonstration that's consistent with the way the world works.

To illustrate, consider the knowledge of the number system that underlies mathematics. Imagine an open stage. A screen comes up to block it, and now one object goes behind the screen, followed by a second object—let's call them Mickey Mouse 1 and Mickey Mouse 2. In our minds, we are representing two Mickey Mouse objects. When the screen is removed, we expect to see two Mickey Mouse objects. If we see three, or if we see just one, it's a violation of our expectations, because nothing was visibly added to or subtracted from what was behind the screen. And indeed, human infants aged about four to five months will look longer when they see such an outcome as opposed to seeing the two objects they apparently expected. My students and I have carried out the same experiment with two nonhuman primate species—

rhesus monkeys living wild on the Puerto Rican island of Cayo Santiago and the cotton-top tamarins in my lab at Harvard—and we have found exactly the same results as the psychologist Karen Wynn found with human infants. These results raise the important question of whether certain aspects of our number faculty—knowledge of number—are innate. This question is important for our understanding of the mechanisms underlying developmental and evolutionary change and for our understanding of the relationship between language and thought. In fact, because animals lack language, studies of their mental representations provide a beautifully clear method for exploring under what conditions language is necessary for thought.

Studies of human infants and animals suggest that evolution endowed these organisms with two core computational mechanisms for number, one enabling precise discrimination of small numbers up to about four and a second enabling large approximate number discrimination. These mechanisms underlie their knowledge of number. What is yet unclear is how these two mechanisms, and perhaps others, work to create a completely different kind of knowledge of number, the kind that underlies adult competence. No animal acquires the integer list that is at the heart of our system of mathematics. This is a statement of the current facts. If true, we need to then ask, Why don't animals and human infants have this system of knowledge? We know that at some point humans are able to do calculus, become bankers, do their taxes—and nonhuman animals are not. What happens in the course of development that separates a human child from a nonhuman animal? By identifying the divergence point, we will show what cognitive ability underlying the adult knowledge of number developed in the child and failed to evolve in nonhuman animals. By identifying similarities as well as differences, we begin to see a pattern of evolution unique in our own species and that of others.

One of the most novel aspects of my work is that unlike researchers who restrict themselves to studies in the wild or in captivity, working on one species, I've taken at least four different approaches to finding out what animals know, think, and represent.

The first is field studies. I go into the wild to understand what kinds of problems shaped the design of the brains of animals in their natural habitat. Watching what animals do tells us what problems their brains need to solve. (The same logic applies to humans, of course, and is one reason why the study of the human mind should not be restricted to studies in the laboratory; we need to figure out what kinds of problems humans confronted in order to understand how our minds were sculpted by environmental forces.) For example, my research in Puerto Rico shows that rhesus monkeys produce different sounding calls for different kinds of food. This suggests not only that they can produce sounds conveying something about their emotions and motivational state as well as about the type or quality of food, but also that they make important discriminations among objects. We can ask how they make such discriminations, how they store this knowledge, and how they acquire it. We can then explore, with experiments designed to suit the behavior of wild animals, how they represent knowledge of food and how they use this knowledge to communicate to others.

So I go to the field, I watch what animals do naturally, and then I come back to the lab, where we have more experimental control, and ask specific questions about their cognitive abilities. In the lab, we note that animals seem to discriminate all sort of objects in their world, and we ask what features are relevant to that kind of discrimination. We now have thirty years of studies showing that animals use tools to extract food from their environment, but what none of these studies have shown is the kinds of representations that the animals bring to the task of tool use. Here's the question: As humans, we know there are certain features of a tool that are relevant to the tool and certain features that are irrelevant. For example, most dishwashers are white, but if we come into the kitchen and see a rainbow-colored dishwasher we don't say, "That's no good. Can't do dirty dishes in that thing." We know that color is irrelevant to whether it's a good dishwasher or a bad dishwasher. When we see animals in the wild—for example, chimpanzees—using stones to crack nuts open, the question then becomes: If we present them with both a stone and a sledgehammer, will they see that the sledgehammer is better designed for the task than the stone? Will they prefer the sledgehammer? Will they realize that if we paint the stone red, it won't make any difference to its functionality? In the lab, we've systematically manipulated all features of the objects, both relevant and irrelevant, to see whether animals make a decision based on those features. We have discovered that animals are in fact quite sensitive to the functionally related features, ignoring differences that have no impact on the task. Their knowledge, in essence, is not a string of associations but a set of principles for organizing different domains of knowledge.

A third step in this program of research is to take these problems to a more neurophysiological level. In collaboration with neuroscientists throughout the United States and internationally, we've begun experiments to look at how the brains of rhesus monkeys, in particular, decode information about their vocalizations. Using recordings from neurons in the various auditory areas of the brain, we play back vocalizations from their repertoire and see how their nervous system decodes that information. This is relatively new work; for a long time now, we've gained an incredible amount of knowledge about the neurobiology of vision using rhesus monkeys as a model, but almost nothing has been done in terms of auditory function. Yet one constraint on our current understanding of the evolution of language and speech is our lack of knowledge about the neurobiology underlying this fantastically complex system. There is a long history of this kind of work with insects, birds, frogs, and bats, but almost nothing on primates, our closest living relatives. Now for the first time we have the tools to probe how the brains of nonhuman primates encode and decode vocalizations.

The fourth step is the comparative studies I alluded to, in which we do the same experiments with animals that we do with human infants, using, for example, the techniques of magical violations to explore the kinds of representations they both bring to the task of enumeration.

Thus, we have a four-pronged approach to understanding the design of animal brains—going from the field back into the lab, then to the neurophysiological level, and finally comparing nonhuman animals with human infants to link developmental processes with evolutionary processes.

With this approach, we can turn to the questions that obsess most of the lay public: Are animals intelligent? Are dogs smarter than cats? Are dolphins smarter than pigeons? Are chimpanzees smarter than dolphins? Are we smarter than these species, and if so, when did we become smarter? These are not good questions. A more productive kind of question is to ask first about the kinds of problems animals confront with respect to survival, and then ask how they solve these problems. What knowledge must they have in order to navigate, mate, win a fight, deceive, learn, communicate, and so on? Each species is intelligent in its own way. The real issue, for me, is not "Are animals intelligent, and do animals think?" but more specific questions, ones we can answer, like: Can animals remember things? And if so, how far back in time can they remember? Do they have memories of what they were like when they were young? Can animals learn about abstract properties of the world, and if so, why would they learn about them? These are questions we can answer with the tools of science. If you then want to say that given those abilities, those animals are intelligent—fine! If you want to say that here are the ways in which animals communicate and it looks like language—that's fine, too. But we should not lose sight of the differences between species, and this includes especially the differences between animals and humans. I do not make this point in order to argue for our uniqueness, but rather to draw attention to the fact that although there are numerous similarities between humans and other animals, the differences are of interest, too, as they point the way to research on the kinds of mechanisms that must have evolved in our past to allow for our particular style of communication, our particular brand of representing the world. Consider, for example, our ability to refer to things in the world: That is, I can talk about a chair, I can talk about my past, I can talk about the future—all in a very abstract way. Do animals have that capacity? If they do, then it resembles a core component of our faculty of language. We can take this general approach and apply it to other faculties or domains of knowledge. We can ask: Do animals have the moral emotions? Can they empathize? Do they feel guilt? Do they feel shame? Are they loyal? Do animals have the capacity for cooperation? Do they engage in reciprocal altruism? These are difficult questions, but we can at least try to make some headway, and in many cases we have made a great deal. So I don't ask, "Do animals think?" I don't ask, "Are

animals intelligent?" I ask questions that have to do with specific cognitive mechanisms that we can identify in humans,
both infants and adults. Similarly, my students and I ask
about how animals solve problems, independently of
whether it is like humans or not. Good biology, as Darwin
highlighted, is comparative biology.

Now, why should we care about such things? There are a
lot of people who love their pets and think their dogs are
Einsteins, and I want to show these people that they should
not be satisfied simply with this intuitive notion. Our intuitions are often not good guides to what animals are doing,
just as our intuitions are often poor guides to how human infants think about the world. One of my goals is to make the
science more palpable and less controversial. People often
present scientists who study animals with incredible observations of what their pets do or don't do. They'll tell you,
"Look, my dog just did the most fantastic thing. I left him six
hours from our house and he found his way home. Isn't that
amazing?" Well, yes and no. No, because it's only one observation and we can't do much with one observation. It's not
that scientists think that any one observation is irrelevant; it's
that one observation is unsatisfying. I want to impress on
people who are interested in animals that they should be un-
satisfied, too. I can give an example of a personal experience
I had with an animal that whetted my appetite for more
questions, and I want the lay public to be equally whetted by
these observations.

I was an undergraduate working at a tourist attraction in
Florida called Monkey Jungle. My job was to feed the monkeys, but I had to make more money, because I was quite
poor, so I decided to take another job raking whatever
dropped below the cages. One day I noticed that a spider
monkey—a species that inhabits the rain forests of South

America—was intently looking at my raking. I didn't think she was that interested in raking, so I thought she might be interested in me. She had a mate who wasn't paying much attention to her. I put the rake down and approached the cage. As I approached, she approached and sat on the other side of the bars from me. She looked me in the eyes and put both of her arms through the cage and wrapped her arms around my neck and cooed. She sat there for quite a long time—a few minutes. Her mate then approached; she let go of me, smacked him in the head, and put her arms around my neck again. You can imagine the thoughts that might go through your head during this experience: You're really connected with the animal. It's in love with you. Or maybe she wanted you to give her more food. Or maybe the previous trainer had trained her to do this. Or maybe she was trying to make her mate jealous—you know, new boy on the block. There were all sorts of possibilities and it's interesting to try to narrow them down. Simple experiments: If somebody else raked the enclosure, would she do the same thing? What if the person doing the raking were a female? What if it were a young boy? What if it were an older man? Those would be the kinds of things you could do in order to eliminate some of the possibilities. If it's specifically me, why me? Is it something about the way I behave? Something about the way I look? Something about the way I smell? Let's change my clothes. Is it just me with certain clothes on? I wore the same clothes every day. Very quickly, you could eliminate a lot of uninteresting possibilities and begin to narrow the question to some interesting possibilities.

Philosophers often use animal examples to show how difficult it is to understand the representations and thoughts of creatures that lack language. Some philosophers claim that in the absence of language there can be no thought. If that's true, we're in a difficult bind when it comes to understanding animal thought, and some would claim that the entire enterprise is bankrupt. Yet there is a long history of research on humans in which tasks have been developed to determine what humans are thinking in the absence of language—a huge amount of work on human infants, who have yet to express their linguistic capacity. What I argue is that some of the most profound problems having to do with the human mind can be addressed only by studying animals. There are three foundations for this claim:

(1) To investigators who hold that a particular kind of thought depends on language, I would argue that the only species you can test that hypothesis on is animals—not human infants, who, although they have yet to develop competence with language, have nonetheless a brain that evolved to be suitable for language and is therefore inappropriate for such a test. Brain-damaged patients who don't have production or comprehension of language are not good subjects, either, because their brains developed with language. If you're interested in the connection between language and thought, you must test that hypothesis on other species. In our lab, and in the field along with scientists like Dorothy Cheney and Robert Seyfarth, we have studied nonhuman primates and other animals to see whether they have a capacity for the kinds of thought that appear to require language. Increasingly there are elegant demonstrations of such representational capacities and thoughts without language.

(2) There are an awful lot of claims about the special nature of particular human thought processes. Beginning in the 1960s, debate focused on the special mechanisms underlying speech. People claimed, for example, that our ability to make categorical distinctions between phonemes, like *ba* and *pa*, was due to one such mechanism. The first refutation of that

idea was made by Patricia Kuhl at the University of Washington, who ran experiments on chinchillas and macaques showing that they have exactly the same perceptual abilities as humans, given the same set of stimuli. Her work has initiated a program of research aimed at identifying whether a particular mechanism is special to humans. The only way to address such claims is by studying animals.

(3) The third reason, more familiar to psychologists and neuroscientists, is the idea that certain kinds of experiments are either unethical or logistically too difficult to run on humans but can be conducted with animals. Although the ethical issue usually dominates this debate, it is equally important to consider the logistics: We may be able to perform better experiments on animals because of the level of control, the types of stimuli presented, and the long-term study of single individuals. Long-term studies of animals, such as Jane Goodall's work on chimpanzees and Cynthia Moss's work on elephants, have provided us with a thirty-year run on the lives of highly social and fascinating creatures. It would be difficult to match such studies with human subjects.

For all these reasons, animal studies are beginning to play a greater role in the cognitive sciences and neurosciences. New techniques allow us to identify animal behaviors that suggest how they think about the world, and the theoretical push we have made is to unite evolutionary theory with modern ideas in cognitive science in a new way. One of the problems with evolutionary psychology is that it has focused exclusively on humans. Broadly defined, evolutionary psychology has been going on since the days of Darwin, who asked questions about the mind with an eye to evolutionary principles. What we're now seeing is an emergence of Darwin's initial intuition—that we can marry evolutionary theory with the cognitive sciences as applied to the study of the animal mind.

We ask questions about the design of the brain, the design of mental states, by looking at how social behavior and ecology shape those processes. For example, we've recently been interested in a domain of knowledge in animals that you might call naïve physics. To what extent do animals make intuitive predictions about physical objects based on the physics of the world? We have devised an experimental procedure modeled on studies performed with human children, in which you drop a ball through an opaque tube with an S-like configuration. Monkeys, and human children, expect the ball to land directly below the release point, not out at the other end of the tube. They seem to be taking gravity into account as a predictive force in their decision, indicating the great difficulty that children and some animals have in suppressing a very strong bias that has been selected for because of the regularities in the world. Gravity is a regularity that all animals on Earth confront. I believe that selection favored brains that innately made predictions about falling objects— and because of this innate sense, it is difficult for animals to override their intuition when the evidence contradicts it.

Why can't animals find the correct location of an object falling through a curved tube? That is, why can't they inhibit their biases and look in a different location? We now know, from studies of the evolution of the brain, that the frontal parts of our brain have undergone extraordinary changes over the past 5 or 6 million years. The frontal region of our brain is roughly 200 percent bigger than that of a nonhuman primate of our size. In humans, that's the part of the brain involved in short-term working memory and where repetitive responses are blocked or inhibited; for example, when we bump into a glass door because we fail to notice that it's closed, we don't repeat that error over and over again. We have a mechanism in the prefrontal region specifically designed to inhibit those kinds of actions—a mechanism that failed to evolve in a significant way in many nonhuman species. What makes this approach to animal studies powerful is that it links to studies of the human brain, creating an intimate connection between thoughts and the neural mechanisms that underlie them.

There are several camps of people who disagree with me, explicitly or implicitly. Those animal investigators trained to a large extent in the Skinnerian tradition find some of the new techniques we're applying to animal cognition sloppy and uninsightful. Then there are people who study human cognition who are becoming converts but find our work annoying because it forces them to rethink their claims about human uniqueness. Another camp works with chimpanzees and doesn't particularly like the fact that the monkeys we study show abilities comparable to chimpanzees. This kind of hierarchical chauvinism continues all the way through the tree of life; there is chauvinism within the community of animal scientists holding that people who work with chimps are doing much more important work than people who work with monkeys.

In the next ten to fifteen years, my hope is that our work, by looking at the problem of cognition from a wide variety of perspectives and different levels of analysis, will demonstrate that interest in the human mind demands an interest in evolutionary theory. It will demonstrate that the theory of evolution leads to novel predictions about the mind, and that we can truly marry studies of animal cognition with the neurosciences. Neuroscientists to a large extent tend to ignore the important variation between species. When they work on rhesus monkeys, for example, they talk about "the monkey."

There are several hundred species of primates, yet neuro-scientists ignore this. Our work will begin to overturn this common and dominant view in the neurosciences. We hope to convince the neuroscience community that variation is wonderful—the truffle of biology, Darwin's truffle. If you're concerned with the design of the mind, the variation between species is of utmost interest. As scientists, we have a common mission: to find out how evolution resulted in different ways of thinking. And by looking at variation, we see natural selection at work, sculpting different kinds of minds.

THE EVOLUTION
OF COOKING

RICHARD WRANGHAM

A lot of people find it hard to live with the idea that we've had a natural history of violence. But if we look at ourselves as an animal, it's clear that natural selection has favored emotions in men that predispose them to enjoy competition, to enjoy subordinating other men, even to enjoy killing other men. These are difficult ideas to accept, and there are people who argue that it's inappropriate to write about such ideas, and they look for ways to undermine the evidence. What they seem to fear is that once a biological component in our violent behavior is recognized, then violence may be seen as inevitable.

RICHARD WRANGHAM is a professor of biological anthropology at Harvard University who studies chimpanzees in Uganda with an eye to illuminating human evolution and behavior. One of Wrangham's central ideas is that we should cherish the parallels between humans and other great apes, because they help us to understand our own behavior. "For all our self-consciousness, we humans continue to follow biological rules," he notes. Wrangham is the author, with Dale Peterson, of *Demonic Males*.

Using biology to analyze human behavior is like going to a psychiatrist and being helped to understand where your behavior is coming from. You're in a little less internal conflict once you can understand what you're doing, and you can shape your own behavior better. But the reaction is not always like that. A lot of people find it hard to live with the idea that our species has had a natural history of violence. But if we look at ourselves as an animal, it's clear that natural selection has favored emotions in men that predispose them to enjoy competition, to enjoy subordinating other men, even to enjoy killing other men. These are difficult ideas to accept, and there are people who argue that it's inappropriate to write about such ideas, and they look for ways to undermine the evidence. What they seem to fear is that once a biological component in our violent behavior is recognized, then violence may be seen as inevitable.

One of the great tenets of behavioral biology in the last three or four decades has been that if you change the conditions that an animal is in, then you change the kind of behavior elicited. What the genetic control of behavior means is not that instincts inevitably pop out regardless of circumstances; instead it is that we're created with a series of emotions that are appropriate for a range of circumstances. The particular emotions that emerge will vary within species but they will also vary with context, and once you know them better, then you can arrange the context. Once you understand and admit that human males in particular have these hideous propensities to get carried away with enthusiasm, to indulge in war, rape, or killing sprees, to get excited about opportunities to engage in violent interactions, then you can start recognizing it and doing something about it. It's better not to wait for experience to tell you that it's a good idea to have a standing army to protect yourself against the neighbors, or that you need to see that women are not exposed to potential rapists. It's better to anticipate these things, recognize the problem, and design protection in advance.

There's still a huge tendency to downplay or simplify gender differences in behavior and emotions. As we achieve a more realistic sense of the way natural selection has shaped our behavior, we'll be increasingly aware of the fact that the emotional responses of men and women to various contexts can be very different. A striking example is the extent to which men and women harbor positive illusions about themselves. In general, women tend to have negative illusions about themselves—that is, they regard themselves as slightly less skilled or competent than they really are. Men tend to have positive illusions: They exaggerate their own abilities, compared with how others see them or how they perform in tests. These tendencies depend a lot on power relations: If you put a woman in a dominant power relation, she tends to have a positive illusion of herself; if you put a man in a subordinate relationship, he tends to have a negative illusion. Nevertheless, the tendencies emerge predictably, and they're dangerous. If you have positive illusions, you think you can fight better than you really can. It looks as though natural selection has favored positive illusions in men because, rather like the long canines on a male baboon, they enable men to fight better against other men who really believe in themselves. You have to believe in yourself to be able to fight effectively; if you don't believe in yourself, others will take advantage of your nervousness and lack of confidence. If you understand something about positive illusions, you can look at an engagement in which each side believes it will win and be a little cynical about it—like a lawyer telling two potential litigants, "Wait a minute, neither of you has got a case quite as good as you think." A more sensitive appreciation for these emotional predispositions generates a more refined approach to violence prevention.

I make my living studying the behavior of chimpanzees in Uganda. I'm interested in looking at the question of human evolution from a behavioral perspective, and I find working with chimps provocative because of the evidence that 5 million, 6 million, maybe even 7 million years ago the ancestor that gave rise to the australopithecines, the group of apes that came out into the savannahs, was probably very much like a chimpanzee. Being with chimpanzees in the forests of Uganda, as in forests anywhere else in Africa, is like entering a time machine; it enables us to think about the basic principles that underlie behavior.

Although humans are enormously different from the apes, the extraordinary thing that has emerged over the last two or three decades—and this has become increasingly clear recently—is that in three big ways in particular, humans are more apelike in their social behavior than you would expect to occur by chance. There's something about our relationship to the apes that has carried through. For example, we know of only two kinds of mammals whose males live in groups of male relatives and occasionally make attacks on individuals in neighboring groups so brutally that they kill them. Those two mammals are humans and chimpanzees. This is odd, and it needs explanation.

Chimpanzees weren't studied in the wild until 1960. Not until fourteen years later were people seeing them at the edges of their ranges; it's just difficult to follow them all over the place. The first brutal attacks were seen in 1974, and these led to the extinction of an entire chimpanzee community in Gombe. People monitored that extinction under Jane Goodall's research direction. And slowly over the years it's been realized that chimps will kill individuals in other communities. We've had chimp killing going on not only in Gombe and at the site I work at—Kibale, in western Uganda—but chimps have also killed other chimps in Budongo, in Uganda and in Mahale, in Tanzania. It just takes time for these observations to accumulate.

Occasionally there's a Julius Caesar-style assassination, which is really intriguing, because it is these tremendously important coalitions within chimpanzee communities that determine a male's ability to do what a male is desperately striving to do all the time, which is to become the alpha male. The question that arises once you see that these coalitions occasionally lead to what are essentially assassinations is, What makes the male alliances so stable normally? How is it that you don't get a constant erosion of confidence? Killings are rare events, but we know a fair amount about them. You can have great imbalances of power—three or four individuals jointly attacking another one, which means that it's essentially safe for the attackers. Various other animals—hyenas, lions, even ants—will also kill rivals in this fashion.

There are three parallels between humans and the great apes that are really striking. The violence that chimps and humans demonstrate is virtually unique to them. Then you have the extraordinary degree of social tolerance in both humans and bonobos, another ape equally closely related to humans. And you have a remarkable degree of eroticism in bonobos, somewhat as in humans. These parallels are not easily explained and raise all sorts of provocative questions, given the fact that humans are so different from the other apes in terms of our ecology, our language, our intelligence—our millions of years of separation.

I've been studying chimps on and off for over thirty years. I began working at Jane Goodall's site at Gombe, which is the archetypal site and represents to many people what the chimpanzee is. In 1984 I moved to Uganda and started work on a forest chimpanzee population, and I began thinking particularly about cultural variation—behavioral traditions— among chimpanzees. One of the lovely things that's going on now is the discovery that in East Africa we have a series of characteristic chimpanzee behaviors different from the behaviors we see in far West Africa—Christophe Boesch's site in the Taï forest of the Ivory Coast, for example, or the Japanese research site in Bossou, Guinea. We're seeing chimp groups in the east that are relatively fragmented, with relatively little sexual activity, with few female/female alliances, and with extreme male dominance over females—all of this different from what's seen in the west. In the most stable groups in the west, the females form alliances, the males are much more respectful of females, and there is much less violence in the community in general. There's much less infanticide and much less severe forms of territoriality. This is exciting, because we can then ask, What are the ecological influences and what are their effects? And what does this mean in terms of trying to reconstruct the kind of chimpanzee that gave rise to us 7 million years ago?

The answers are becoming clearer. In my field work, I am trying to understand what it is about the ecology that leads to differences in behavior. A key factor that has been given little attention is the fact that in some populations the apes are able to walk and feed at the same time. In others they're not, because there's no food for them as they're walking. This sounds like a trivial difference, but it seems to be enormously important, because if you can walk and feed at the same time, then you can stay in a group with your friends and relations without additional members intensifying the feeding competition. On the other hand, if you are walking without feeding between the food patches, then every time another chimp comes along and joins your party, feeding competition is intensified in these food patches, and there is no amelioration when you're moving between food patches. The long-term effect of this is that it fragments the parties, and it's the fragmented nature of these parties of chimps that cannot walk and feed at the same time that underlies all of the social differences.

There are two fascinating things about human evolution that we have yet to fully come to grips with. One is the evolution of cooking. Whenever cooking happened, it must have had tremendous effects on us, because cooking enormously increases the quality of the food we eat and the range of food items we can eat. We all know that food quality and food abundance are key variables in understanding animal ecology. But the amazing thing is that although there is no conventional wisdom about when cooking evolved, social anthropology and all sorts of conventional wisdom tell us that humans are the animals that cook. We distinguish ourselves from the rest of the world because the rest of the world eats raw stuff and we eat cooked stuff. The best anthropology can do at the moment is to say that maybe sometime around 250,000 or 300,000 years ago cooking really got going, because there's excellent archeological evidence of earth ovens in that period.

This is fine, but long before earth ovens came along we must have learned to cook. You would expect cooking to be associated with evidence in the body of food being easier to digest—such as smaller teeth, or maybe a reduction in the size of the rib cage as the size of the stomach got smaller, or maybe a reduction in the size of the jaw. And there's one point in human evolution when all of that happened: 1.9 million years ago with the evolution of the genus *Homo*. That is where we must look for evidence that cooking was adopted.

Once cooking happens, it completely changes the way the animal exploits its environment. Instead of moving from food patch to food patch, and eating as it goes or eating in the food patches, now for the first time it has to accumulate food, put it somewhere, and sit with it until it's cooked. That might take twenty minutes; it might take half an hour; it might take several hours. The effect is that all of a sudden there's a stealable food patch. Once you have a stealable food patch, life being what it is, somebody's going to come along and try to steal it. What this means is that you have a producer/scrounger dynamic, in which you've got individuals producing and individuals scrounging—and, horribly, females were the producers and males were the scroungers.

Once you've got males bigger than females—they were 50 percent bigger by the time we're talking about—then the effects on the social system would be large.

What we've got to think about is the idea that once you have females ready to make a meal by collecting food and cooking it, then they're vulnerable to having their food taken away by the scroungers—the big males—who find it easier not to go and collect food themselves or cook it but just take it once it's ready. Therefore the females need to make protective alliances in order to protect themselves from thieving males, and this is the origin of human male/female relationships. The evolution of cooking is a huge topic that is virtually completely neglected. And whatever view you take of cooking, you have to say that it's a problem needing to be addressed.

The second problem is this: There is, in a number of ways in the evolution of humans, evidence of our behaving and looking as if we had the characteristics of a juvenile animal. For 100 years or more, people have talked about the idea that humans might be a pedomorphic species—a species that shows juvenile characteristics—but this is too global a way to think about it. Still, it remains the case that much in our behavior, when compared with that of our closest relatives, looks more playful and less aggressive when you're thinking about interactions at a social level within a group. We are also more sexual and more ready to learn—characteristics generally associated with juvenility.

In a fascinating parallel, the bonobos—the second in the great pair of our two closest relatives—show all sorts of traits that are pedomorphic. We can see this in the head, where the morphology of the skull looks like that of an early adolescent or late-juvenile chimpanzee, and much bonobo behavior looks juvenile. Bonobos are more playful, they're less sex-differentiated in all sorts of aspects of their behavior, they're more sexual, and so on. We've yet to pin down where this pedomorphic change has come from and what it means.

We've already got some wonderful examples of similar phenomena in other animals in the context of domestication. When we look at the differences between wolves and dogs, for example, we see remarkable parallels to the differences between chimpanzees and bonobos. In each case, for a given size of animal, you have the skull being reduced in size, and the components of the skull being reduced in size, including the jaws and teeth, and the skull looking more like the skull of a juvenile in the other form. The dog's skull looks like that of a juvenile wolf, and the bonobo's skull looks like that of a juvenile chimpanzee. And the behavior of each of them seems to have strong components of the juvenile of the other species.

This leads to the thought that species can self-domesticate. There is good reason to think that the bonobos evolved from a chimpanzeelike ancestor as a consequence of being in an environment where aggression was less beneficial and selection favored individuals that were less aggressive. Over time, selection built on those slight variations in the timing of the arrival of aggressive characteristics in adult males. It was constantly pushing back, favoring individuals that retained more juvenile-like behavior—and even juvenile-like heads, because the brain is what controls behavior. Later, what you had was a species that had effectively been tamed—been self-domesticated.

There is experimental evidence of this process. The Russian geneticist Belyaev, for example, took wild foxes and selected them purely for tameness. Foxes are ready to breed at the age of eight months, so Belyaev was able to see the results relatively quickly. After only twenty-five generations, he found not only that the descendant foxes were as tame as dogs but also that they had a series of characteristics that seemed to have come along for the ride—incidental consequences that were not selected for but just evolved anyway. There were dramatic morphological ones—like the star mutation, the white spot on the forehead you see in horses and cows and goats—which are evidently associated genetically with tameness, for reasons that are completely mysterious. Other morphological changes—like curly hair, short tails, lopped ears—occur in a number of domesticated animals. Why these correlated effects occur we don't know.

In addition, you get smaller brains. This is a remarkable thing about human evolution. We tend to think there has been a continuous rise in human brain size over the last 2 million years, but actually, over the last 30,000 years, brain size has decreased by 10 to 15 percent. The standard explanation for this is that we became more gracile at the same time—that we became thinner-boned—which meant that we were lighter in body weight, and because there tends to be a correlation between body weight and brain weight, this explains our smaller brains. But I don't see any reason why brain size should be correlated with the amount of meat we carry on our bodies. This gracility is exactly the same pattern we see in the evolution of dogs from wolves, or bonobos from chimpanzees, or domesticated foxes from wild foxes. In all these cases, an increasing gracility of the bone is an incidental effect.

I think we have to start entertaining the idea that we humans in the last 30,000, 40,000, or 50,000 years have been domesticating ourselves. If we're following the bonobo or dog pattern, we're moving toward a form of ourselves with more and more juvenile behavior. Once you start thinking in those terms, you realize that we're still moving fast. Tooth size, for example, is strongly genetically controlled and develops with little environmental influence—and it is continuing to decline fast. Current evidence indicates that we're in the middle of an evolutionary event in which tooth size is decreasing, jaw size is decreasing, brain size is decreasing, and it's quite reasonable to imagine that we're continuing to tame ourselves. The way it's happening is the way it's probably happened since we became permanently settled in villages 20,000 or 30,000 years ago, or earlier.

People who are antisocial, for example, have their breeding opportunities reduced. They may be executed, they may be imprisoned, or they may be punished so badly that they're kept out of the breeding pool. Just as there is selection for tameness in the domestication process of wild animals, or just as in bonobos there was a natural selection against aggressiveness, there's a sort of social selection against excessively aggressive people within communities. Human beings appear to be becoming, increasingly, a peaceful form of a more aggressive ancestor.

THE COMPUTATIONAL PERSPECTIVE

DANIEL C. DENNETT

When I go to a workshop or conference and give a talk, I'm actually doing research, because the howls and screeches and frowns that I get from people, the way in which they react to what I suggest, is often diagnostic of how they are picturing the problems in their own minds. And in fact people have very different covert images about what the mind is and how the mind works. The trick is to expose these images, to bring them up into public view and then correct them. That's what I specialize in.

DANIEL C. DENNETT is University Professor, professor of philosophy, and director of the Center for Cognitive Studies at Tufts University. A philosopher by training, he is known as the leading proponent of the computational model of the mind. He is the author of *Content and Consciousness; Brainstorms; Elbow Room; The Intentional Stance; Consciousness Explained; Darwin's Dangerous Idea; Kinds of Minds; Brainchildren; Freedom Evolves;* and *Breaking the Spell.* With Douglas Hofstadter, he coedited *The Mind's I,* and he is the author of over 200 scholarly articles on various aspects of the mind.

If you go back 20 years, or if you go back 200 years, 300 years, you see that there was one family of phenomena that people just had no clue about, and those were mental phenomena—that is, the very idea of thinking, perception, dreaming, sensing. We didn't have any model for how that was done physically at all. Descartes and Leibniz, great scientists in their own right, simply drew a blank when it came to trying to figure these things out. And it's only really with the ideas of computation that we now have some clear and manageable ideas about what could possibly be going on. We don't have the right story yet, but we've got some good ideas. At least you can now see how the job can be done.

Coming to understand our own understanding, and seeing what kinds of parts it can be made of, is one of the great breakthroughs in the history of human understanding. If you compare it with, say, our understanding of life itself or reproduction and growth, those were deep and mysterious processes 100 years ago and forever before that. Now we have a pretty clear idea of how things reproduce, how they grow, repair themselves, fuel themselves. All of those formerly mysterious phenomena are falling into place.

And when you look at such phenomena, you see that at a very fundamental level they're computational—that is, there are algorithms for growth, development, and reproduction. The central binding idea is that you can put together not billions but trillions of moving parts and get entirely novel, emergent, higher-level effects, and the best explanation for what governs those effects is at the level of software, the level of algorithms. If you want to understand how orderly development, growth, and cognition take place, you need to have a high-level understanding of how those billions or trillions of pieces interact with each other.

We never had the tools before to understand what happens when you put a trillion cells together and let them interact. Now we're getting those tools; even the lowly laptop gives us hints, because we see phenomena happening on our desks that would astound Newton or Descartes—or Darwin, for that matter. Phenomena that look like sheer magic. We know they aren't magic. There's not a thing that's magical about a computer. One of the most brilliant things about a computer is that there's nothing up its sleeve. We know to a moral certainty that there are no morphic resonances, psyonic waves, spooky interactions; it's good old push/pull, traditional, material causation. And when you put it together by the trillions with software, you get all this magic that's not really magic.

The idea of computation is a murky one; it's a mistake to think we have a clear, unified, unproblematic concept of what counts as computation. It's less clearly defined than the idea of matter or the ideas of energy or time in physics, for instance. Even computer scientists have only a fuzzy grip on what they actually mean by computation. The question is where to draw the line between what is computation and what isn't. That's not so clear. But that doesn't mean we can't have good theories of computation. Almost any process can be interpreted through the lens of computational ideas, and usually it's a fruitful exercise of reinterpretation. We can see features of the phenomenon through that lens that are essentially invisible through any other.

Human culture is the environment we live in. There's the brute physical environment—the streets, the air we breathe, the water we drink, the cars we travel in—and then there's all the communication going on around us in many different media: everyday conversation, newspapers, books, radio, television, the Internet. Pigeons live in our world, too, but they're oblivious to most of it; they don't care what's written in the newspaper they find their crumbs on. It's immaterial to them what the content, what the information, is. For us it's different; the information is really important.

If we think about the informational world our species lives in, we see that in fact it's got a lot of structure. It's not amorphous. Everything is not connected to everything else. There are lots of barriers. There's an architecture to this world of communication, and that architecture is changing rapidly, in ways we don't understand yet.

Let me give you a simple example of this. You could tune in the Super Bowl a couple of years ago and see these dot-com companies pouring an embarrassingly large amount of their initial capitalization into one ad for the Super Bowl; they were trying to get jump-started with this ad. And this was curious. If this was an Internet company, why didn't it use the Internet? Why do this retrograde thing of advertising on regular broadcast television? And the answer, of course, is that there is a fundamental difference in the conceptual architecture of those two media. When you watch the Super Bowl, you're part of a large simultaneous community—and you know it. You know that you are one of millions, hundreds of millions of people. You're all having the same experience at once, and you know that you are. And it's that second fact—it's that reflexive fact—that's so important. You go to a Web site and there might be 100 million people looking at that Web site but you don't know that. You may have read that somewhere . . . but you're not sure, you don't know. The sense you have when you're communing on the Web is a much more private sense than when you're watching something on network television. And this has huge ramifications regarding credibility. An ad that works well on television falls flat on the Web, because the people who see it, read it, listen to it, don't know what audience they're a part of. They don't know how big a room they're in. Is this a private communication or a public communication? We don't know yet what kind of fragmentation of the world's audiences is going to be occasioned by the Internet. The Internet brings people together, but it also isolates them in a way we haven't begun to assess. That sense of being utterly lost that neophytes have when they first get on the Web—choosing search engines, knowing what to trust, where home is, whom to believe, what sites to go to—arises because everybody is thirsting for reliable informants, signposts.

This geography of available information was established over centuries in the traditional media. You went to the *Times* and you read something there, and it had a certain authority for you. Or you went to the public library and read something in the *Encyclopedia Britannica*. These institutions had their own character, their own reputations, and their reputations were shared communally. It was important that your friends also knew that the *Times* or the *Encyclopedia Britannica* was an important place to look. Suppose somebody writes and publishes a volume called "Sammy's Encyclopedia of the World's Information"? It might well be the best encyclopedia in the world, but if people in general don't realize it, nobody's going to trust what's in there. It's this credibility issue that, as far as I can see, has not yet even begun to crystallize on the Web. We're entering uncharted waters there, and the outcome is hard to predict.

Human experience changed tremendously in the last century—and especially over the last decade. For instance, I'd guess that the average western-world teenager has heard more professionally played music than Mozart ever heard in his whole life (not counting his own playing and composing and rehearsal time). It used to be that hearing professional musicians in performance was a very special thing. Now *not* hearing professional musicians is a very special thing—there's a soundtrack almost everywhere you go. This is a huge change in the auditory structure of the world we live in. The other arts are similarly positioned. There was a time when just seeing written words was a rarity. Now everything has words on it. People can stand in the shower and read the back of the shampoo bottle. We are completely surrounded by the technology of communication—and that's new. Our species has no adaptations for it, so we're winging it.

There are lots of patterns in the world. Some are governed by the law of gravity, some by other physical principles. And some of them are governed by software. That is to say, the robustness of the pattern—the fact that it's salient, that you can identify it, that it keeps reproducing itself, that it can be found here, there, and elsewhere, and that you can predict it—is not because there's a fundamental law like the law of gravity that governs it but because these are patterns that occur wherever you have organisms that process information. They preserve, restore, and repair the patterns and keep them going. And that's a fundamental, new feature of the universe. If you went to a lifeless planet and surveyed all the patterns on it, these patterns wouldn't be there. They're the patterns you can find in DNA—those are the ur-patterns, the ones that make all the other patterns possible. They're also the patterns you find in texts. They have to have some physical embodiment in nucleotides or ink marks or particles and charges. But what explains their very existence in the universe is computation, the algorithmic quality of all things that reproduce and have meaning, and that make meaning.

These patterns are not, in one sense, reducible to the laws of physics, although they are based in physical reality. The explanation of why the patterns form as they do has to go on at a higher level. Douglas Hofstadter once offered a very elegant simple example: We come across a computer and it's chugging along, chugging along, chugging along. Why doesn't it stop? What fact explains the fact that this particular computer doesn't stop? And in Hofstadter's example, the reason it doesn't stop is that pi is an irrational number. What? Well, pi is an irrational number, which means that it's a never-ending decimal, and this particular computer program is generating the decimal expansion of pi, a process that will never stop. Of course, the computer may break. Somebody may come along with an ax and cut the power cord—but as long as it keeps powered, it's going to go on generating these digits forever. That's a simple concrete fact that can be detected in the world, the explanation of which cites an abstract mathematical fact.

Now, there are many other patterns in the world that aren't as arcane and have to do with the meaning we attach to things. Why is he blushing? There's a perfectly good explanation of what the *process* of blushing is: Blushing is the suffusion of blood through the skin of the face. But *why* is he blushing? He's blushing because he thinks she knows some fact about him that he wishes she didn't know. That's a complex, higher-order, intentional state, one that's visible only when you go to the higher, intentional level. You can't see that by looking at the individual states of the neurons in his brain. You have to go to the level at which you're talking about what this man knows and believes and wants.

The intentional level is what I call the "intentional stance." It's a strategy you can try whenever you're confronted with something complex in nature. It doesn't always work. The idea is to interpret that complexity as one or more intelligent, rational agents that have agendas, beliefs, and desires and are interacting. When you go up to the intentional level, you discover patterns that are highly predictive, robust, and not reducible in any meaningful sense to the lower-level patterns at the physical level. In between the intentional stance and the "physical stance" is what I call the "design stance." That's the level of software.

The idea of abstraction has been around for a long time, and 200 years ago you could enliven a philosophical imagination by asking what Mozart's Haffner Symphony is made of. It's ink on pieces of paper. It's a sequence of sounds as played by people with various stringed instruments and other instruments. It's an abstract thing. It's a symphony. Stradivarius made violins; Mozart made symphonies, which depend on a physical realization but don't depend on any particular one. They have an independent existence, which can shift from one medium to another and back.

We've had that idea for a long time, but we've recently become much more comfortable with it, living as we do in a world of abstract artifacts that jump promiscuously from medium to medium. It's no longer a big deal to go from the score, to the music you hear live, to the recorded version of the music. You can jump back and forth between media very rapidly now. It's become a fact of life. It used to be hard work to get things from one form to another; that's not hard work anymore; it's automatic. You eliminate the middleman. You no longer have to have the musician to read the score, to produce the music. This removal of all the hard work in translating from one medium to another makes it all the more natural to populate your world with abstractions, because you find it's hard to keep track of what medium they're in. And that doesn't matter much anymore—you're interested in the abstraction, not the medium. Where did you get that software? Did you go to a store and buy an actual CD and put it in your computer, or did you just download it off the Web? It's the same software, one way or another. It doesn't really matter. This idea of medium neutrality is one of the essential ideas of software, or of algorithms in general. And it's one that we're becoming familiar with, but it's amazing to me how much resistance there still is to this idea.

An algorithm is an abstract process that can be defined over a finite set of fundamental procedures—an instruction set. It is a structured array of such procedures. That's a very generous notion of an algorithm—more generous than many mathematicians would like, because I would include by that definition algorithms that may be in some ways defective. Consider your laptop. There's an instruction set for that laptop consisting of all the basic things its CPU can do; each basic operation has a digital name or code, and every time that bit sequence occurs, the CPU tries to execute that operation.

You can take any bit sequence at all and feed it to your laptop as if it were a program. Almost certainly, any sequence that isn't *designed* to be a program to run on that laptop won't do anything at all—it'll just crash. Still, there's utility in thinking that *any* sequence of instructions, however buggy, however stupid, however pointless, should be considered an algorithm, because one person's buggy, dumb sequence is another person's useful device for some weird purpose, and we don't want to prejudge that question. (Maybe that "nonsense" was included *in order* to get the laptop to crash at just the point it crashed!) One can define a more proper algorithm as one that runs without crashing. The only trouble is that if you define algorithms that way, then probably you don't have any on your laptop, because there's almost certainly a way to make almost every program on your laptop crash. You just haven't found it yet. Bug-free software is an ideal that's almost never achieved.

Looking at the world as if everything were a computational process is becoming fashionable. Here one encounters not an issue of fact but an issue of strategy. The question isn't "What's the truth?" The question is "What's the most fruitful strategy?" You don't want to abandon standards and count everything as computational, because then the idea loses its sense; it doesn't have any grip anymore. How do you deal with that? One way is to try to define, in a rigid centralist way, some threshold that has to be passed, and refuse to call a process computational unless it has properties A, B, C, D, and E. You can do that any number of ways and it will save you the embarrassment of having to say that everything is computational. The trouble is that anything you choose as a set of defining conditions will be too rigid. There will be processes that meet those conditions but aren't interestingly computational by anybody's standards, and there are processes that don't meet the standards but nevertheless are significantly like the things you want to consider computational. So how do you deal with the issue of definition? By ignoring it—that's my suggestion. Same as with life! You don't want to argue about whether viruses are alive or not; in some ways they're alive, in some ways they're not. Some processes are obviously computational. Others are obviously not computational. Where does the computational perspective illuminate? Well, that depends on who's looking at the illumination.

I describe three stances for looking at reality: the physical stance, the design stance, and the intentional stance. The physical stance is where the physicists are; it's matter and motion. The design stance is where you start looking at the software—at the patterns that are maintained—because these are designed things that are fending off their own dissolution. That is to say, they are bulwarks against the Second Law of Thermodynamics. This applies to all living things and also to all artifacts. Above that is the intentional stance, which is the way we treat that specific set of organisms and artifacts that are themselves rational information-processing agents. In some sense, from the intentional stance you can treat Mother Nature—that is, the whole process of evolution by natural selection—as an agent, but we understand that that's a *façon de parler*, a useful shortcut for getting at features of the design processes that are unfolding over eons of time. Once we get to the intentional stance, we have rational agents, we have minds, creators, authors, inventors, discoverers—and everyday folks—interacting on the basis of their take on the world.

Is there anything above that? Well, in one sense there is. People—or persons, as moral agents—are a specialized subset of the intentional systems. All animals are intentional systems. *Parts* of you are intentional systems. You're made up of lots of lesser intentional systems—homunculi of sorts—but unless you've got multiple personality disorder, there's only one person there. A person is a moral agent—not just a cognitive agent, not just a rational agent, but a moral agent. This is the highest level I can make sense of. Why it exists at all, how it exists, the conditions for its maintenance: these are very interesting problems. We can look at game theory as applied to the growth of trees—they compete for sunlight, it's a game in which there are winners and losers.

But when we look at game theory as applied not just to rational agents but to people with a moral outlook, we see some important differences. People have free will; trees don't. It's not an issue for trees in the way it is for people.

What I like about the idea that people are animals with free will is that it agrees with philosophical tradition (including Aristotle and Descartes, for instance) in maintaining that people *are* different—that people aren't *just* animals. Traditional theorists completely disagree, of course, on what that difference consists of. Although it's a naturalization of the idea of people, it does say they're different, and this, I discover, is the thing that most entices and upsets people about my view. There are those who want people to be more different than I'm allowing. They want people to have souls, to be Cartesian people. And there are those who are afraid that I'm trying to differentiate people too much from the other animals with my claim that human beings really are, because of culture, an importantly different sort of thing. Some scientists view this claim with skepticism, as if I'm trying to salvage for philosophy something that should fall to science. But in fact my view about what's different about people is a scientific theory; it stands or falls as an implication of a scientific theory, in any case.

Regarding my own role in cognitive science—as to whether I consider myself a philosopher or a scientist—I think I'm good at discovering the blockades of imagination, the bad habits of thought, that infect how theorists think about the problem of consciousness. When I go to a workshop or conference and give a talk, I'm actually doing research, because the howls and screeches and frowns that I get from people, the way in which they react to what I suggest is often diagnostic of how they are picturing the problems in their own minds. And in fact people have very different covert images about what the mind is and how the mind works. The trick is to expose these images, to bring them up into public view and then correct them. That's what I specialize in.

My demolition of the Cartesian theater, of Cartesian materialism, is just one of these campaigns of exposure. People often pay lip service to the idea that there isn't any privileged medium in the brain playing the role that Descartes assigned to the nonphysical mind as the theater of consciousness. Nevertheless, if you look closely at what they are thinking and saying, their views make sense only if you interpret them as covertly presupposing a Cartesian theater somewhere in their model. Teasing this out, bringing it to the surface and then showing what you might replace it with is, to me, very interesting work. Happily, some people have come to appreciate this as a valuable service that somebody like me, a philosopher, can perform: getting them to confront the hidden assumptions in their own thinking and showing how those hidden assumptions blind them to opportunities for explaining what they want to explain.

WHAT SHAPE ARE A GERMAN SHEPHERD'S EARS?

STEPHEN M. KOSSLYN

There is a gigantic project, yet to be done, that will root psychology in the rest of natural science. Once this is accomplished, you'll be able to go from phenomenology (things like mental imagery) to information processing . . . to the brain . . . down through the workings of the neurons, including the biochemistry, all the way to the biophysics and the way that genes are up-regulated and down-regulated. This is going to happen; I have no doubt at all. When it does, we're going to have a vastly better understanding of human nature than at any other time in human history.

STEPHEN M. KOSSLYN, the John Lindsley Professor of Psychology at Harvard University, has published over 250 papers on the nature of visual mental imagery and related topics. He is a cofounder and senior editor of the *Journal of Cognitive Neuroscience* and has served on several National Research Council committees advising the government on new technologies. His books include *Image and Mind; Ghosts in the Mind's Machine; Elements of Graph Design; Wet Mind;* (with Olivier Koenig); *Image and Brain;* and *Psychology;* (with Robin Rosenberg).

———————

For the last thirty years, I've been obsessed with a question: What shape are a German shepherd's ears? Of course, I'm not really interested in that question specifically; if I were, I could just go out and look at dogs. What I'm really interested in is how people answer the question from memory. Most people report that they visualize the dog's head and mentally "look at" its ears. But what does it mean to visualize something? What does it mean to "look at" something in your mind? There's no little person in your mind actually looking at a picture. If there were, there would have to be a little person inside that person's head, and so on and so on, and that doesn't make any sense.

For many years, we tried to collect objective evidence showing that when you have the experience of visualizing, there's actually something pictorial in your head. There are parts of the brain that are physically organized so that when you look at something, a corresponding pattern is physically laid out on the cortex. Even if your eyes are closed when you visualize, the first visual area in the processing stream is often activated during visual imagery; moreover, the way it's activated depends on what you're visualizing. If you visualize something vertical, there is activation along the so-called vertical meridian; if you visualize something horizontal, the activation flips over on its side. Similarly, visualizing objects of different sizes changes the pattern of activation in ways very much like what occurs if you look at objects of the corresponding sizes.

But I've been working on answering this question—not about the dog but the question behind that question, what imagery is—for some thirty years now, and I want to move on. Instead of just trying to establish that there are actual mental images and that these are bona fide representations that have a functional role in processing systems, I want to ask, "So what? Who cares?" Lately I've been working on something I've tentatively called the Reality Simulation Principle. It's built on my lab's findings that most—about two-thirds—of the same parts of the brain are involved both in visual mental imagery and in visual perception. That's a huge amount of overlap, which leads us to suspect that the mental image of an object can have the same impact on the mind and body that seeing the actual object would have. My notion is that once the brain systems are engaged, they don't know (so to speak) where the impetus came from; they can produce the same effects whether you activated the process endogenously (from information in memory) or exogenously (from looking at something).

The Reality Simulation Principle describes how to use mental images as stand-ins for actual objects—basically, how to manipulate yourself. It's useful to understand the principle in conjunction with what I call the GITI cycle, which stands for Generate, Inspect, Transform, Inspect. If mental images can simulate actual objects and scenes, you can generate the image, inspect what you've got, transform it, and inspect the result. This can be done iteratively, meaning that you can take advantage of the Reality Simulation Principle to do all sorts of good things for yourself.

What kinds of good things am I talking about? Memory is one obvious example. From the work of the cognitive psychologist Alan Paivio and countless others, we know that you're able to remember objects better than pictures of objects, and pictures of objects better than words. It also turns out that if you visualize the objects named by words, you do better in memory tests than you would otherwise. Consequently, we're now interested in topics such as hypnosis. We can hypnotize you and have you visualize an object and imagine that it's actually a three-dimensional object appearing in glorious vivid detail. In this case, we expect that your memory will be boosted even further.

Neuroscientists such as Marc Jeannerod and Jean Decety have shown that imagining doing something recruits most of the brain mechanisms that would guide the corresponding actual movements. And people in sports psychology have shown that by imagining that you're engaging in some activity, you'll get better at actually doing it. This process, too, involves generating an image, inspecting the image, transforming it by imagining your movements, "seeing" what the result would be, and then cycling through again. The next time through, you can change the image, depending on the result you "saw." If you imagine you're playing golf, for example, and your ball didn't go in the hole, you can imagine what would happen if you putted a little more gently. Mental practice clearly works. By understanding how the mechanisms of imagery work, we can optimize this mental practice.

The Reality Simulation Principle can also be used to acquire self-knowledge. Try this one out: Imagine it's dusk, you're walking alone, you're late. You start to walk faster and then notice a shortcut through an alley. It's getting darker but you really don't want to be too late, so you start toward it. Then you notice three guys lingering near the mouth of the alley, smoking cigarettes. Now think about a first scenario: The three guys appear to be in their early twenties; they're wearing long droopy shorts, dirty T-shirts, and baseball caps on backwards. As you get close, they stop talking and all three heads swivel and start tracking you. How do you feel?

Now try the same thing, except make them three balding, middle-aged, overweight accountants wearing suits. They're standing there smoking cigarettes, and all three heads swivel and start tracking you. How do you feel now?

What if the guys are black or Latino? How do you feel? If you can sort out your own emotional landscape by this mental simulation, you may well discover things about yourself that will surprise you. Make those middle-aged accountants black and see how you feel. Some people who confront their reactions to the simulations may find that what they thought were racial issues are actually class issues. These kinds of simulations can produce self-knowledge and help you improve your emotional intelligence.

You can also manipulate your body with imagery. Obviously, when you have a sexual fantasy that's what you're doing. And if you imagine something scary—an anticipated encounter with an authority figure, say, or a walk along a narrow, crumbling path on the side of a mountain—your palms will begin to sweat and your heartbeat will increase. Clearly mental imagery affects the body, but I'm thinking of something more interesting than what's evident in these examples. One of the effects we're studying now is how to change your hormonal landscape by manipulating your images. There's the so-called Victory Effect: If you're a male and you win a contest, your testosterone goes up, and if you lose, it goes down. This is perhaps not a surprise, but it also turns out that if you watch your favorite team win, your testosterone goes up, and if your team loses, it goes down. This works even if you're watching a chess tournament, so it's not about being aroused.

Why is this of interest? It turns out that spatial abilities in men vary with their testosterone levels. Much research suggests that the relation between testosterone levels and spatial abilities is a U-shaped function; your spatial abilities are not as good if you have too much or too little testosterone. As you get older, your testosterone levels and your spatial abilities both drop. There's a lot of evidence that the two are connected. The question is, Can you manipulate your testosterone levels—and thereby manipulate your spatial abilities—by running imagery simulations, watching yourself win or lose? If the Reality Simulation Principle is correct, you can. This is work-in-progress in my lab, in collaboration with Peter Ellison and Carole Hooven. Stay tuned.

My point is that you can use the Reality Simulation Principle in lot of different ways, including some that are not intuitively obvious, such as manipulating your hormonal landscape. Mental imagery is also important in creativity and problem solving. Einstein reported that most of his thinking was accomplished with the aid of mental images, prior to any kind of verbal or mathematical statement. We know quite a bit now about how to use images in the service of solving problems and being creative. People also claim that you can manipulate your health by using what I'm calling the Reality Simulation Principle. I'm a little skeptical about that. It's certainly the case that you can manipulate the placebo effect to some extent, but the medical effects of the Reality Simulation Principle are probably not huge. If perceptual events have no effect, then you shouldn't expect imagery to have an effect. Watching a particular event doesn't seem to help cure cancer, which makes me think that imagery won't, either.

In trying to understand mental imagery, my premise is that "the mind is what the brain does." Of course, that's a little too glib. Really, the mind is what the cortex does, since the brain also does things that aren't mental, such as respiration. If that's the case, then the question becomes, How do we understand information processing in the brain? This is one of the deepest questions in psychology, and probably in science in general. It's really a mystery. How is it that semantics and the meaning of things can dictate a sequence of events in this wet machine? The wet machine itself has some 100 billion neurons, each of which, on average, has 10,000 connections. Sure, it's complicated, but ultimately you can understand the brain in terms of chemistry and physics.

But how does this machine produce semantically interpretable, coherent sequences of activity and allow these activities to be modulated by the semantics of what it registers from the world? When you say something to me, it's not just sound patterns—the *content* of what you say influences what my brain is doing. How I'm going to respond is a consequence of how my brain processes the input. The only way I know of even to begin to think about such questions is to think about how the brain processes information, how it "computes." Think for a moment about physical events such as the status of bytes in a computer. Each bit in each sequence of eight bits is either on or off. You can physically describe the nature of this machine and the hardware, but you can also think about representation: What does that pattern of physical activity stand for? You can think about interpreted rule-based systems, where the representations have an impact on other parts of a system, causing other representations to be formed, modified, combined, or operated on in various ways, and causing outputs to be generated. In this regard, it's useful to think about computation in a computer to describe how the mind works, even though it's the wrong metaphor for a brain.

A computer is based on a Von Neumann architecture, in which you have a strict separation between memory and the central processing unit. This means that there is a strict separation between operations and representations, which sit passively in memory. The CPU is essentially a switching device that uses instructions to dictate what it's going to do, both in terms of how it interprets successive sets of instructions and what it does with the representations. The very idea of representation depends on how the CPU is set. That is, the exact same pattern of bytes can represent a number, a letter, or part of a picture, depending on how it's being interpreted. Once an operation is performed, the results go back into memory and serve as input for additional processes. The computer is useful as a way of thinking about all of this, but it's not a model of how the brain works; the brain doesn't work like this at all. But using computation as a model for understanding the brain lets us appreciate the exquisite dance of events at different levels of analysis. It's a wonderful mystery. How can an idea arise from wet stuff? How can an idea influence what's going on in the wet stuff?

We—fortunately!—don't need to answer such questions to make progress in understanding the mind. My work has been strongly influenced by the computational perspective, but I think the important part is what's been found, the empirical discoveries. When I was a graduate student, I stumbled onto the basic phenomenon I've been studying for thirty-odd years now. In my first year of graduate school at Stanford— this was 1970—studies of semantic memory were really hot. Allan Collins and Ross Quillian published a simulation model in 1969 in which they claimed that information is stored in long-term memory in the most efficient way possible. (This makes no sense for the brain, by the way, since storage space is apparently not an issue, although it is in a computer.) They posited that memories are organized into hierarchies in which you store information in as general a representation as possible. For example, under "animals," you've got a representation of animals in general, and then birds, mammals, reptiles, and so on. And under "birds," you have canaries, robins, and so on. The notion was that you store the various properties as high up in the hierarchy as you can, rather than redundantly duplicating them. For example, birds "eat," but so do lizards and dogs, so we store this property higher up, with the concept of animals. You tag the exceptions (such as the fact that, unlike most birds, ostriches don't fly) in a lower level.

One way to test this theory is to record response times. If you present people with a statement such as "A canary can sing" and ask them to decide whether it is true or false, the information required to make the decision should be stored in the same place; that is, "canary" and "sing" should be bound together, at a low level in the hierarchy. But if you ask them to evaluate "A canary can eat," the participants should have to traverse the network to find a connection between the two concepts, if "eat" is stored with "animal." Thus, evaluating this statement should take a little longer than evaluating "A canary can sing"—and it does! Unfortunately for the model, though, distance in a semantic net turned out not to be crucial. My first-year project at Stanford showed that the response time is simply due to how closely associated the terms are, not to the distance between them in a semantic net. The theory was appealing, but the data were easily explained with a mundane idea. What's the quote—something about a beautiful theory killed by ugly facts. Well, that was it.

However, the story doesn't end here. In one of the experiments I asked people to respond to the statement "A flea can bite—true or false?" Two people in a row said, "False," and afterward I asked them why. One said that he had "looked for" a mouth and couldn't find one. The other said he had "looked for" teeth and couldn't "see" any. This idea of "looking for" and "seeing" didn't fit in at all with Collins and Quillian's network-based computer model, so I started thinking about that. My idea was that perhaps some of the participants had used mental imagery to evaluate the statements, and if so then their response times should reflect properties of the image—not distance in a semantic net, association strength, or anything like that. So, I telephoned everybody I'd already tested and asked them if they had tended to visualize when they were answering the question. Roughly half said they had and half said they hadn't. I plotted the data separately for the two groups. And *voilà!* For the people who reported that they used imagery, how strongly associated the properties were with the animals had nothing to do with how quickly they responded. For those people, the crucial variable was the size of the properties: the larger the property, the faster they could "see" it.

I immediately designed an experiment in which I pitted the two characteristics—strength of association and size—

against each other. For example, I asked people to decide whether statements such as "A mouse has whiskers" were true or false. The trick here was that they considered traits that are small and highly associated (such as whiskers for a mouse), or traits that are large and not highly associated (such as a back for a mouse), or traits that the animal doesn't have at all (such as wings for a mouse). I found that if I instructed people to visualize, the critical factor was how big the property was: The bigger it was, the faster the responses were. If I asked them not to visualize but to answer intuitively as fast as they could, the pattern reversed. In this case, the response speed depended on how associated the trait was, not how big it was.

The next question was how to think about those results. Fortuitously, while I was doing these experiments I was also taking a computer programming class. This was in the days when you used punch cards. You had to go to the computer center, submit your stack of cards, and stand around looking at a monitor, waiting for your job to come up so you could see whether it bombed, which you could tell by how long it ran. One of the exercises in the class was to program a set of little subroutines that generated geometric shapes—triangles, squares, circles—and adjust how big they were and where they were positioned. You had to do things like make a Christmas tree by recursively calling the same routine, generating a triangle and plotting the triangle at different sizes in different positions, overlapping them to produce the design.

As I was doing this, it suddenly occurred to me that this was an interesting model of mental imagery. We can think of imagery as having four main components: a deep representation, which is an abstract representation in long-term memory; a surface representation, which is like a display in a cathode-ray tube; generative processes between the two, so that the surface geometry is reconstructed in the "mental display" on the basis of the deep representation; and, finally, interpretative processes that run off the surface image, interpreting the patterns as representing objects, parts, or characteristics.

This metaphor was neat and led me to conduct a lot of fruitful research. In fact, my first dozen papers or so were largely a result of following up implications of this metaphor. But it had a fundamental drawback: It was a metaphor, not an actual theory. No matter how hard you hit somebody in the head, you're not going to hear the sound of breaking glass—there's no actual cathode-ray tube screen in the head. Even if there were, we would just be back to the problem of needing a "little person" in the head to look at the screen (and another little person in his head, and so on and so on). This immediately led me to start thinking about how to program a system in which there are arrays that function as a buffer, and a surface image is created by positioning points in the array to depict shapes. If this pattern of points was the surface image, and the array was a short-term memory buffer, then you could have a much more abstract, languagelike representation that's actually stored in long-term memory—which could be operated on to create the image. The neat idea here, I thought, was that you could have your cake and eat it, too: What's stored is abstract, but this can be used to create something very concrete and picturelike.

One of the virtues of the computer analogy is that it focuses you on the idea of processing systems—not just isolated representations or processes but sets of representations and processes working together. Nobody had ever tried to work out in detail what a processing system that uses images would look like. In fact, the few detailed models of imagery that existed all focused on specific artificial tasks and tried to model them using standard list structures. There were no images in the early computer-based models of imagery. We decided to take seriously the idea that perhaps mental images aren't represented in the same way as language; perhaps they really *are* images. Steve Schwartz and I built a series of simulation models showing that such an approach is not only possible but accounts for much of the data. We published our first paper on this in 1977, and another in 1978. I also wrote a book on it in 1980, called *Image and Mind*, in which I worked the idea out in much more detail than anyone ever cared about. As far as I can tell, neither the model nor the book had much impact. I remember asking one of my professors at Stanford for his opinion of the model, and he thought it was too detailed. Psychologists generally don't like having to work with a really detailed theoretical framework, and that was basically the end of it. I have a mild frontal-lobe disorder that leads me to perseverate, and thus I've continued to work out the theory and do experiments anyway. My 1994 book on imagery, *Image and Brain*, is a direct outgrowth of the earlier work but maps it into the brain. The Europeans—especially the French—and the Japanese seem interested, if not the Americans.

That said, I should note that lately there are signs that interest in mental imagery is picking up. This might be a result of another round in my old debate with Zenon Pylyshyn. He's a good friend of Jerry Fodor, but, unlike Fodor, Pylyshyn has maintained forever that the experience of mental images is like heat thrown off by a light bulb when you're reading: It's epiphenomenal; it plays no functional role in the process. Pylyshyn believes that mental images are just languagelike representations and that it's an illusion that there's something different about them. He published his first paper in 1973. Jim Pomerantz and I replied to it in 1977, and the debate has been rolling along ever since.

Pylyshyn has great distain for neuroscience, to put it mildly. He thinks it's useless and has no bearing at all on the mind. I really don't know what brings him to this conclusion. I suspect it's because he is one of the few people (less than 2 percent of the population) who does not experience imagery. He apparently doesn't even get jokes that depend on images. He also probably rejects the very idea of imagery on the basis of his intuitions about computation, based on Von Neumann architecture. He's clearly aware that computers don't need pictorial depictive representations. His intuitions about the mind may be similar. But this is all speculation.

Pylyshyn is not only against theories that are rooted in neural mechanisms—he thinks theories of the logical structure of language should be a model for all other types of theories—but he's also against neural network computational models. I've probably published eight to ten papers using network models. At one point in my career, I worked on the nature of spatial relations. I had the idea that there are actually two ways to represent spatial relations among objects. One is what I call categorical, in which a category defines an equivalence class. Some examples of such categorical spatial relations would be "left of," "right of," "above," "below," "inside," and "outside." If you are sitting across from me, from your point of view this fist is to the right of this open palm, and that's true for all these different positions [moving his hand about, always to the right of the vertical axis created by his fist]. "Right of" defines a category, and even though I move my hand around, all these positions are treated as equivalent. This is useful for doing something such as recognizing a human form, since the categorical spatial relations among my forearms and upper arms, lower legs and thighs, head and neck, neck and body, and so on, do not change. Parts that are "connected to" (another categorical spatial relation) other parts remain so, no matter how I contort my body. Descriptions of arrangements of parts using categorical spatial relations are handy for recognizing objects, because if you store a literal picture in memory, an upright posture might match well, but if I bend over and try to touch my toes, the resulting image would not match well.

But categorical spatial relations are not useful at all for reaching or navigating. Just knowing that this fist is to the left of this palm won't allow me to touch it precisely; I've got to know its exact location in space. If I'm walking around the room and all I know is that the table's in front of me, that's not helpful, because "in front of" is a categorical relation and thus is true for an infinite number of relative positions. Not good enough for navigating. Thus, I posit a second type of spatial relation, which I call coordinate: Relative to an origin, the metric distance and direction are specified.

In my lab, we have shown that the left cerebral hemisphere is better at encoding categorical spatial relations— which makes sense, because categories are often language-based. The right hemisphere is better at encoding coordinate spatial relations—which makes sense, because navigation is performed better by this hemisphere. We've constructed a whole raft of neural network models showing that if you split a model—a network—into two separate streams, one for each type of representation, it does better than if you have a single system trying to make both the categorical and coordinate representations. The point is not so much that the hemispheres are different but that the brain relies on two distinct ways to code spatial relations. This claim caused a minicontroversy. I was delighted to see not long ago in the *Journal of Cognitive Neuroscience* that researchers—whom I didn't know at the time—tested over 100 people after they turned off one hemisphere at a time for medical reasons, and showed that with challenging tasks where you have to make categorical versus coordinate spatial relations judgments, the laterality effects I predicted worked beautifully.

This is really just one little corner of what I do, and ultimately is related to my imagery work. I've always argued that imagery has to be understood in a system that includes languagelike propositional representations as well as depictive representations. I don't think of the mind as purely imaginal. That can't be true. It's got to depend on coordinating many different types of representations that interact in intricate and interesting ways. The distinction between the two types of spatial representations invites a further distinction between different forms of imagery that make use of the different sorts of spatial relations. And in fact we have evidence for such a distinction. One clear conclusion from all this: Imagery isn't just one thing.

Why is the system that underlies imagery organized the way it is? Good question. One way to approach this fundamental concern has been articulated by Dan Dennett, Steve Pinker, and their colleagues. These theorists are trying to cash out the evolutionary psychology program. Instead of thinking about behaviors as the products of evolution, they are thinking about how the modular structure of information processing in the brain is a consequence of evolution. That's an interesting program, and I think it has a bright future. But at this juncture I'm a little queasy about the fact that this enterprise is not particularly empirical. Science is the process of finding things out. You've got to do studies to find things out. It's helpful to have theories as a base from which you can direct your attention to issues and questions, but then you've got to do the actual research.

If you asked me to explain the direction of mind science

writ large, I'd say that what you're going to see is a bridging between cognitive neuroscience—where the mind is conceived of as what the brain does—and genetics. Those are the two really hot areas right now, and there's a giant gulf between them.

When I was writing an introductory psychology textbook, I read a lot of behavioral genetics. I was struck by the fact that these people are trying to bridge the gap from genes to behavior in one fell swoop, and they're not doing that good a job at it. They're not doing that well in linkage studies that try to connect variability in a behavior with variability in different types of alleles. Sometimes they manage 2 percent of the variance. It occurred to me that they're leaving out the middleman. They want to think in terms of the model: genes —> behavior. But it would be much better to think in terms of: genes —> brain, and then brain —> behavior. That is, genes influence behavior and cognition via what they do to the brain. Thinking about this has gotten me very interested in genetics, but not in the sense that genetics is a blueprint. Most genes functioning in the adult brain seem to be up-regulated and down-regulated by circumstances. They turn on and off.

Here's an example that illustrates the general point (developed by the psychiatrist Steven Hyman, who happens to be the current provost of Harvard): If you want to build muscles, you lift weights. If the weights are heavy enough, they will damage your muscles. That damage creates a chemical cascade, reaching into the nuclei of your muscle cells and turning on genes that make proteins and build up muscle fibers. Those genes are turned on only in response to the environmental challenge. That's why you've got to keep lifting heavier and heavier weights. The phrase "No pain, no gain" is literally true in this case. Interaction with the environment turns on certain genes that otherwise wouldn't be turned on; in fact, they'll be turned off if certain challenges aren't faced. The same is true in the brain. Growing new dendritic spines, or even replenishing neuro-transmitters, is linked to genes' being turned on and off in response to the brain, which in turn is responding to environmental challenges.

I'm absolutely fascinated by the really Big Question, how genes allow the brain to respond to the tasks at hand. When genes are turned on and off, this affects what neu-rons are doing, which then of course affects how blood is allocated, which in turn affects cognition and behavior. There is a gigantic project, yet to be done, that will root psychology in the rest of natural science. Once this is ac-complished, you'll be able to go from phenomenology (phenomena such as mental imagery) to information pro-cessing (phenomena you can model on the computer) to the brain. We'll understand how particular kinds of infor-mation processing arise in the brain, down through the workings of the neurons, including the biochemistry, all the way to the biophysics and the way that genes are up-regu-lated and down-regulated.

This is going to happen. I have no doubt at all. When it does, we're going to have a vastly better understanding of human nature than at any other time in human history. If you want to understand evolution, the products of evolu-tion ultimately are genes. Why not study the genes, if you want to understand the reasons behind the brain's organiza-tion? There are reasons we have those genes rather than other ones; that's where the evolutionary story comes in. But my particular brain or your particular brain is the way it is not only because of the particular genes we have but also because of the way the environment up-regulated or down-regulated those genes during development, sculpting our brains certain ways, and because of the ways our genes respond to environmental and endogenous challenges. All of this is empirically tractable. The tools are available, the questions are clear, and we know what sort of answers to seek. Time to get cracking!

DARWIN Y
LA TERCERA CULTURA

Lee Smolin,
Marc D. Hauser, and
Robert Trivers

. . . there is a deep relation between Einstein's notion that everything is just a network of relations and Darwin's notion because what is an ecological community but a network of individuals and species in relationships which evolve? There's no need in the modern way of talking about biology for any absolute concepts for any things that were always true and will always be true.

—Lee Smolin

What I'm interested in is how science can fuse with and energize moral philosophy to create some powerful new ideas and findings at the interface. This is not to say that science will take over philosophy. If this new enterprise works at all it will be through a deep collaboration, working to find out the origins of our moral judgments and how they figure in our ethical decisions and moral institutions.

—Marc D. Hauser

I believe that self-deception evolves in the service of deceit. That is, that the major function of self-deception is to better deceive others. Both make it harder for others to detect your deception, and also allow you to deceive with less immediate cognitive cost. So if I'm lying to you now about something you actually care about, you might pay attention to my shifty eyes if I'm consciously lying, or the quality of my voice, or some other behavioral cue that's associated with conscious knowledge of deception and nervousness about being detected. But if I'm unaware of the fact that I'm lying to you, those avenues of detection will be unavailable to you.

—Robert Trivers

L ast year *Edge* received an invitation from Juan Insua, the director of Kosmpolis, a traditional literary festival in Barcelona, to stage an event at Kosmopolis '05 as part of an overall program "that ranges from the lasting light of Cervantes to the (ambiguous) crisis of the book format, from a literary mapping of Barcelona's Raval district to the dilemma raised by the influence of the Internet in the kitchen of writing, from the emergence of a new third culture humanism to the diverse practices that position literature at the core of urban creativity."

Something radically new is in the air: new ways of understanding physical systems, new focuses that lead to our questioning of many of our foundations. A realistic biology of the mind, advances in physics, information technology, genetics, neurobiology, engineering, the chemistry of materials: all are questions of capital importance with respect to what it means to be human.

Charles Darwin's ideas on evolution through natural selection are central to many of these scientific advances. Lee Smolin, a theoretical physicist, Marc D. Hauser, a cognitive neuroscientist, and Robert Trivers, an evolutionary biologist, traveled to Barcelona last October to explain how the common thread of Darwinian evolution has led them to new advances in their respective fields.

The evening was presented by Eduard Punset, host of the internationally-viewed Spanish-language science television program Redes, and a best-selling author in Spain. A Redes television program based on the event was broadcast throughout Spain and Latin America.

The house was packed. The Barcelona press was present and covered the event with headlines and cover stories in the major media.

COSMOLOGICAL EVOLUTION: LEE SMOLIN

LEE SMOLIN, a theoretical physicist, is the founding member and research physicist at the Perimeter
Institute in Waterloo Canada. He is the author of
*The Life of The Cosmos; Three Roads to Quantum Grav-
ity;* and the recently published *The Trouble With
Physics: The Rise of String Theory, The Fall of a Science,
and What Comes Next.*

———

I'm a theoretical physicist, and I'm here to talk about natural
selection and Darwin. The main thing that I'll have to
communicate is why somebody who thinks about what the
laws of nature are has something to say about Darwin and
the impact and the role of Darwin in our contemporary
thinking over all fields.

But as a way of getting into that, I want to say something—and you're going to hate me for this—about John.
When John talks about the Third Culture, what he has
done, besides create the idea, is create a group of people.
I don't know if it's a community, there are very close
friendships in it, many of which, I believe, are people who
meet through John. This conversation that he talks about
is a real conversation, which, at least as far as I know, was
not happening and would not be happening were it not
for John.

I think it is not so usual that John appears with the people whom he talks about and writes about, and it is wonderful to be here with him. So I wanted to say thank you publicly because when I started to think about Darwinism, I didn't know any biologists and I didn't know any psychologists—the academic world is very narrow—and I didn't know any artists or digerati. It is because of this involvement that I met many of the people I admired and whose work has inspired me. Now I want to make a claim, and my claim is that while Darwin's ideas are certainly completely absorbed and verified within biology, the whole impact of Darwin's ideas is still yet to be absorbed and felt, and the impact is going to happen in my field of theoretical physics and cosmology and I see it happening in other fields—mathematics, social fields, and so forth. Also I want to make a hypothesis about why—and I'll speak about my field because I don't have any rights to speak about another field—but I think the resonances and the similarities are there.

In my field, two things happened in the 20th century that we're absorbing. One of them was Einstein and the revolution of physics started by Einstein, both relativity theory and quantum theory. And I'm going to claim in my brief time—I'm not going to have time to fully justify—that the main development and the main meaning of Einstein's contribution are closely related to the main meaning of Darwin's contribution. At least I'll say why in a few minutes.

The other thing that's happened in my field has been accelerating really for the last five years. I can't see the audience, so I don't know if any of my friends who are physicists in Barcelona are here, but I think they will agree that a very strange thing has happened to our field, which is that we used to think that the purpose of theoretical physics was to understand what the laws of nature are—to learn the laws of nature—and we're not done with that. But what we've discovered on the way is that we really have to answer a different question—and for our field a very new question—which is: Why these laws and not other laws?

I don't actually believe it's going to work all the way through, but the most successful approach to putting all the laws together and unifying physics is string theory. In the last few years, we've learned that there are an infinite number of these theories, and the best we can do as far as looking for a unified theory so far is to have an infinite list of theories, one of which might describe our universe. So the question has gone from, what are the laws, to why these laws and not other laws? Now I think that the only rational way to approach that question is through Darwin's thinking, that is, through evolution by natural selection.

If I had been an educated person rather than a narrowly-educated person in science, I would have known the quotation that I'm going to read to you, which is from the 1890s by the American philosopher Charles Pierce, who was one of the founders of the philosophical school called "pragmatism." Already in the 1890s he was worrying about the question of why these laws, which shows that sometimes philosophers really are a century ahead of the scientists.

He wrote—and I'll read it slowly so that a translator can get it: "To suppose universal laws of nature capable of being apprehended by the mind and yet having no reason for their special forms but standing inexplicable and irrational is hardly a justifiable position." He's saying, it's not enough to know what the laws are, you want to ask why these laws, and just to say "these are the laws, tough," is irrational and unjustifiable. He says: "Uniformities are precisely the sorts of facts that need to be accounted for. Law is par excellence the thing that wants a reason." And now here is his thesis: "The only possible way of accounting for the laws of nature, and for uniformity in general, is to suppose them the results of evolution." By which, from the context, we know it's evolution by natural selection because he was fully absorbing the impact of Darwinism and that's a lot of what his philosophy, and the American Pragmatists' philosophy, was about.

In my own work, I began to worry about this problem about fifteen years ago—why these laws and not other laws—and I went looking for a method to attack that problem because there's another side to it, which is that the laws we happen to have are very special. The laws we happen to have have a number of free constants that can be freely adjusted, and about twenty-five years ago, Martin Rees and colleagues—these are great cosmologists and astrophysicists—began to realize that if you varied these numbers—these numbers refer to things like the mass of the elementary particles, the mass of the electron, the mass of the proton, the strengths of the different forces—the world we live in would fall apart.

Imagine that the universe can be set up by a dial—by a machine with a set of dials where you dial these constants. If you go away—in any direction—from the settings of the dials that we have, there are no more stars, there are no more hundred-something nuclei which are stable, there is just hydrogen. There's no structure, there's no energy, the world is just dead internal equilibrium.

So the fact that we live in a world which is as complex as it is, which has stars that live for billions of years, which enables life to evolve on planets, which is a process that takes hundreds of millions to billions of years, is due to these constants being finely-tuned—the dials being precisely tuned. They were worried by that and most of the people who found it are sort of liberal British Anglicans, and they have an answer that vaguely has something to do with God, or is something which is logically equivalent to God. And I was disturbed by that, and was looking for an alternative which would be a scientific explanation of how the dials got turned.

At about that time, somebody gave me a book by Richard Dawkins and I started to read it and it opened up my eyes to the kinds of explanations which are possible in biology. I copied it and I made a little cosmological theory that I don't have time to tell you about, but I might, in the discussion, discuss the way in which these dials get tuned by a process which is just like natural selection.

It works better than the theory that it was made by God or is logically equivalent to made-by-God in that—and I think that this is characteristic of biology and Darwinian thought/Darwinism—the process of natural selection produces not just what we see, but a whole very complicated set of interrelations among the different species and among the individuals of the species which leads to predictions that these guys can test. Similarly, the style of Darwinian thought and cosmology and physics has led to predictions that we could test. That impressed me very deeply and as I started to look into it more, I began to see a connection with what really was the field I was trained in, which is relativity and quantum theory.

Roughly speaking the connection is the following—and I'm just going to say some key words and define them, and some key statements, and then, if people want, I can elaborate on it. What did Einstein do, in one sentence? Before Einstein—and what this has to do with is the nature of space and the nature of time—physics, which was based on Newton's physics, was formulated the following way: there's a fixed absolute space, it's eternal, it goes on forever, it was always there, and particles come and move around in this space, and they have all their properties defined with respect to that space.

The space never changes. Similarly, time is absolute, flows whether anything is happening or not, the same way: In a certain sense, space and time are outside the phenomena that we observe, and prior to it. And Newton believed that for a good reason—that space was really God's way of sensing his creation. These were really theological ideas for Newton and they became how people did science.

Einstein replaced that with another idea, which is much more commonsense, which is that space is a system of relations amongst things in the world. Where this pen is in space is not some absolute thing that only God can see, it's where it is relative to the glass, the bottle. So space has no meaning apart from a network of relationships, and time is nothing but change in that network of relationships. And that was an idea that some philosophers—of course we scientists don't pay attention to philosophers, I said—but some philosophers, like Leibnitz and Mach had been arguing for against the success of Newton's physics.

But it was Einstein who first took these ideas and made them into science, and made them into science which, as far as we know, is true and is much better science than the previous Newton science. So the changes from a world in which things exist against an absolute preexisting framework, to a world which is nothing but a network of relations, where change is nothing but change in those relations.

Now here's something that's fascinating: We draw pictures in our work—when I work on the theory of space-time and quantum space—we draw pictures which are networks of relations and how they change in time, and our pictures look just like pictures of ecological networks that these people study. Or the Internet. Or networks of people in interaction, in social interaction. And we began to notice that. Why do our pictures look the same as these pictures from biology and social theory and the Internet and so forth? I think the reason is that there is a deep relation between Einstein's notion that everything is just a network of relations and Darwin's notion because what is an ecological community but a network of individuals and species in relationships which evolve? Theres no need in the modern way of talking about biology for any absolute concepts for any things that were always true and will always be true.

That's what I think is important about Darwin and, again, why I think that it's closely related to Einstein's ideas. It's just the start of what I hope is a conversation.

Let me close by saying what the scariest idea for me is, because these are really revolutionary ideas and that means that they're scary to those of us who think about them. We look forward to the generation to whom they will not be scary, which will mean that the revolution is over and we can go and have fun—not that we don't have fun, but they can take over.

The scary thing is that if the laws evolve, what does that really mean? If what we're used to thinking of as laws which are absolutely true, true for all time—the phrase "God-given" comes to mind because that's how the founders of modern physics like Newton thought about them. If laws instead become, as Pierce said, explainable through a process of evolution, then that means time is very real, in a way that it is not in other representations of physics.

But it's also very scary because we're used to thinking of laws as absolute, and if laws evolve, then at least I and the people I work with get very confused. What does it mean? Is there nothing? What is guiding the evolution? Are there just other laws, which you guys don't have to worry about because you have our laws to hold things steady, but when our laws start to evolve, is there anything under anything/everything? Or is it possible that people in the future, when this revolution that Einstein started is over, will be perfectly comfortable living in a world described by the philosopher Pierce in which there is nothing to laws but this temporary momentary result of an on-going process of evolution?

MORAL MINDS:
MARC D. HAUSER

MARC D. HAUSER is an evolutionary psychologist and biologist, a Harvard College professor of psychology, biological anthropology, and organismic and evolutionary biology, and the director of the Cognitive Evolution Laboratory. He is the author of The Evolution of Communication; Wild Minds; and the recently published Moral Minds: How Nature Designed Our Universal Sense of Right and Wrong.

———————

I want to echo one of Lee's comments about John and say thanks for a slightly different but related reason. What I believe John has allowed many of us to do, which is exciting, is to communicate our passion to a broader audience, escaping academia to exchange with interested professionals and others from a broader slice of mental life. This not only enriches understanding at a broader level, but also allows for a more interesting dialogue. So thank you, John.

Today, I want to engage you in a game that I hope will bring to life my thinking in the last few years. Here is the game: I want you to turn to your neighbor and pair up into a team—okay, you're a pair now. Please pair off with somebody. One of you will be designated the donor in this game, and the other person is the receiver. Please chose a role, either donor or receiver. Please pair off, as I really need your data; I'm an experimentalist. Okay, here is the game. It's going to be played once. I'm giving—play along with me—each donor ten euros. The game starts in the following way: the donor is going to turn to the receiver and offer some proportion of that ten euros—one, two, up to ten. The receiver will respond by either accepting the offer or rejecting it. If the receiver accepts, he or she gets what was offered and the donor gets what's left; if the receiver rejects, nobody gets any money. So now, donor, make an offer to the receiver, and receiver, respond.

Okay. Let me collect some of the data by asking you to raise your in hands in the following way: Of the donors, how many offered between one and three euros? Raise your hands. How many offered between four and six euros? How many offered between seven and ten? Only a few very generous people, and most of you offered in the four to six range. Now, how many of the receivers rejected their offers? Keep your hands up—of those with your hands up, how many of you got offers of one to three euros? One to three euros? Small numbers? Small offer? How much were you offered? Uno. Okay, good. Now what I want you to do with me is think through the logic of the game as if you were an economist. If you were trying to maximize your returns, donors should have given the lowest offers possible, and they should have been thinking that the receivers should accept any offer, because one euro is certainly better than zero euros. You didn't have anything to begin with, so one is better than nothing, two is better than nothing, and so is three.

But it turns out that when this game is played, in many, many different countries, the typical offer is exactly in the range seen here: about four, five, or six euros—it's much more than if you were trying to maximize your own returns. And yet we seem to make this calculation very quickly, spontaneously, almost without thinking. That's example number one. Keep in mind.

Here's example number two: I want you to imagine that you are watching a train moving down a track, out of control. It's lost its brakes. If the train continues, it will hit and kill five people. But you are standing next to the train tracks, and you can flip a switch and turn that train onto a sidetrack, where there's one person. Now the train will kill that one person. Here's the question: Is it permissible—morally permissible—for you to flip the switch, causing the train to kill one but save five? If you think yes, raise your hand. If you think no, raise your hand. Okay, most of you think it is permissible.

Now, second example: Here comes that train again; it's going to kill the five if it keeps going. You are standing next to a very heavy, fat person, and you can throw them onto the tracks, killing them, but the train will stop before the five. Is it morally permissible to throw the fat person? Yes? We've lost half of you! Or more. Okay, what happened? Why do so many of you switch from a permissible to a forbidden judgment?

Here is the idea that I want to give you tonight, in the next few minutes: There has been a long history—a very old tradition—about the sources of our moral judgments. Where do they come from? Many moral philosophers, legal scholars, think that the way that we deliver a moral judgment, like you just did, comes from reasoning. It comes from thinking about the principles, maybe utilitarian (more saved is better than less saved). You work through the principles in a conscious, reasonable, rational way. This was certainly a view that someone like Kant was very much in favor of: how you deliberate with your moral judgments. Now opposing that view—diametrically opposed—is a view that dates back at least to Hume, which is that when we give a moral judgment, we do so based on our emotions. It just feels wrong, or it feels right, to do something, and that's why we do it, that's why we say it's morally right or morally wrong.

What I want to argue for you today is that both of these views, which have dominated the entire field of moral philosophy, are wrong, at least in one particular way. What you just did tonight is an example of why it's wrong. You delivered those moral judgments quickly, probably without reasoning, and without consciously thinking about principles. And if I were to ask you, as I have asked literally thousands of people—on the Internet, in small-scale societies like the Mayans and hunter-gatherers of Africa—people deliver exactly the same judgments that you did tonight but are incapable of justifying why. Typically they say it's a hunch or a gut feeling. So for example, let me illustrate by telling you about my father's response to these cases. He was a distinguished physicist. But I am not picking on the physicists. When I first presented him with these moral dilemmas, the ones you just answered, he said, yes, you can flip the switch, turning the train onto the sidetrack; he said, yes, you can push the fat man onto the train. I said, but Dad, really? He answered, "Of course, it's still five versus one." He was following good utilitarian guidelines.

And now I give him case number three.

You are a doctor in a hospital and there are five people in critical care. Each person needs an organ to survive. The nurse comes to the doctor and says, "Doctor, there's a man who has just walked into the hospital, completely healthy, coming in for a visit. We can take his organs and save the five."

"Can you do that, Dad?"

He immediately replies, "No, you can't just kill somebody!"

I then say, "But you killed the fat man five seconds ago."

He then volleys back: "Okay, you can't kill the fat man."

"But what about the switch?" I say.

Defeated, he replies, "Okay, not the switch either." And the whole thing unravels because there is not a consciously accessible set of principles that people can recall and use to justify what's going on. And it's not based on emotion. It's based on a calculus that the mind has, that it evolved to solve particular kinds of moral dilemmas. And it's not learned, either; it's there in place early in development.

If I have been sufficiently clear this far, you may have already figured out where I am going and the connections I wish to make with another discipline. The core of my argument for moral judgment derives from an argument that the linguist Noam Chomsky developed almost fifty years ago concerning the nature of language, its representation in the mind, and its normal functioning in every human. The idea here in a nutshell is that the way our moral sense works is very much like the way language works. There is a universal set of moral principles that allows the establishment of a set of possible moral systems. In this sense, perhaps this provides some convergence with what Lee said just a few seconds ago. In the same way that you might want to ask about possible universes, I want to ask the question about possible moral systems—that the mind is constraining the range of possible variation.

So the deep aspect of Chomsky's thinking about language, which I think is directly translatable into the way we think about morality and the way we do the science, is by imagining that humans are equipped with, born with, a set of universal principles. What culture can do is change things locally, like a parameter; there are switches. Once you turn something on, things can change.

Let me try to give you a concrete example of some work that a student of mine just recently did. There's a population of people in Panama, Central America, called the Kuna Indians. There is one part of their range that is very remote, and this is where we worked. They live in a quite simple type of society, including small-scale agriculture and fishing. We went there recently and gave them moral dilemmas exactly like the ones you just answered. They weren't about trolleys; they were about wild animals. So in one example, there are crocodiles coming to eat five people in the river; you're in a canoe, you can move those crocodiles off to where they will kill one. Is it permissible? The Kuna said it was, virtually every single person we asked. Here's the second case: you can throw somebody into the river so that the crocodiles will eat him, and save the five. Is that permissible? No. They're showing the same parallel system of psychology where an intended harm—using someone as a means to a greater good—is less permissible than a foreseen consequence that causes the same harm. So in the switch case of a train, you foresee the consequence, but you are not intending the harm as a means to the greater good. The Kuna are sensitive to this distinction, but here's where the cultural aspects move in to make this case more interesting: the Kuna Indians are much more willing to say that it's permissible to throw the fat man in front of the crocodiles than we are in our society. They have an unstated policy—a social behavior of high levels of infanticide. Killing, as a part of society, is much more common. And that's the way in which culture can potentially change the dynamics of how the judgment gets made. In other words, we will see a universal principle such as the means-versus-side-effect distinction, but culture can change how much more impermissible the means-based harm is when contrasted with the foreseen side effect.

At this point this is still looking relatively abstract and theoretical and what I'm interested in is how science can fuse with and energize moral philosophy to create some powerful new ideas and findings at the interface. This is not to say that science will take over philosophy. If this new enterprise works at all it will be through a deep collaboration, working to find out the origins of our moral judgments and how they figure in our ethical decisions and moral institutions. Let me end with a few more cases to make this all a bit more concrete.

Consider a disorder that people are aware of, acutely aware of, in many societies. It's called psychopathy—people who are known for massive killings. They kill, often with no regret: They don't feel guilt, they don't feel shame, and they don't feel empathy. Now people have described that as a problem of lacking any moral sense. I think that's completely the wrong interpretation. What psychopathy is, is a case where they have completely intact moral knowledge. They would judge the cases I just gave you like everybody else in the room. What makes them moral monsters is that they lack the kinds of emotions that we have to prevent them from doing horrible things. They don't have braking emotions. In this view, emotions don't dictate our moral judgments, but they do guide our moral behavior, how we act. We are now engaged in a collaborative project, working to actually test psychopaths, to see whether that is in fact the case. It is too early to tell, but stay tuned.

The second part of the story is to use, as John mentioned a few minutes ago, some of the modern techniques in the neurosciences where you can image the brain, attempting to understand which parts of the brain are active, how they are engaged when we come to our moral judgments, and how they resolve conflict.

To sum up, we're in an extremely exciting phase now, where a set of questions that were forever the providence of moral philosophy and law are now coming directly into contact with the sciences. This is exciting because both areas are working together, and it may have direct implications for the law and the extent to which formal institutions like law and religion penetrate our evolved moral sense.

DECEIT AND SELF-DECEPTION: ROBERT TRIVERS

ROBERT TRIVERS' scientific work has concentrated on two areas, social theory based on natural selection and the biology of selfish genetic elements. He is the author of Social Evolution, Natural Selection and Social Theory: Selected Papers of Robert Trivers; and coauthor (with Austin Burt) of Genes in Conflict: *The Biology of Selfish Genetic Elements*. He was cited in a special *Time* issue as one of the 100 greatest thinkers and scientists of the 20th Century.

Why do I talk about, or wish to talk about, deception and self-deception in the same breath? Because I think you miss the truth about each if you are not conscious of the other and the relationship between the two. If by deception you only think of conscious deception, where you're planning to lie or aware of the fact that you're lying, you will miss all the lying that goes on that the individual is unaware of, and this may be the larger portion of lies and deception that is going on.

Conversely, if you think about self-deception without comprehending its connection with deception, then I think you'll miss the major function of self-deception. In particular, you'll be tempted to go the route that psychology went a hundred years ago or so and think of self-deception as defensive: I'm defending my tender ego, I'm defending my weak psyche. And you will not see the offensive characteristic of self-deception.

What do I mean by that? I mean that I believe that self-deception evolves in the service of deceit. That is, that the major function of self-deception is to better deceive others. Both make it harder for others to detect your deception, and also allow you to deceive with less immediate cognitive cost. So if I'm lying to you now about something you actually care about, you might pay attention to my shifty eyes if I'm consciously lying, or the quality of my voice, or some other behavioral cue that's associated with conscious knowledge of deception and nervousness about being detected. But if I'm unaware of the fact that I'm lying to you, those avenues of detection will be unavailable to you.

Regarding the second argument, it is intrinsically difficult, and mentally demanding, to lie and be conscious about it. The more complex in detail the lie—the longer you have to keep it up—the more costly cognitively. I believe that selection favors rendering a portion of the lie unconscious, or much of the knowledge of it unconscious, so as to reduce the immediate cognitive cost. That is, with self-deception you'll perform better cognitively on unrelated tasks that you might have to do moments later than if you had just undergone a lot of consciously mediated deception.

Let me step back and say a word or two about the underlying logic. First of all, we understand that if we are making an evolutionary argument in terms of natural selection, we are talking about benefits to individuals in terms of the propagation of their own genes, and there are innumerable opportunities in nature to gain a benefit by deceiving another.

However, the reverse is true for the deceived.

The deceived is typically losing knowledge or resources or whatever, resulting in a decrease in the propagation of their genes. So you have what we call a co-evolutionary struggle, with natural selection improving deception on the one hand, and improving the ability to spot deception on the other.

Now let me just say that deception is a very deep feature of nature. At all levels, all interactions, e.g., viruses and bacteria, often use deception to get inside you. They may mimic your own cell surface proteins. They may have other tricks to deceive your system into not recognizing them as alien and worthy of attack. Even genes inside yourself, which propagate themselves selfishly during meiosis, may do so by mimicking particular sub-sections of other genes so as to get copied an extra time, even though the rest of the genome, if you asked their opinion, would be against this extra copying.

When you turn to insects and larger creatures like those, we know that in relations between species, again there's a huge and rich world of deception. Considering insects alone: They will mimic harmless objects so as to avoid detection by their predators. Or they will mimic poisonous or distasteful objects to avoid being eaten. Or they will mimic a predator of their predator, so as to frighten away their predator. Or, in one case, they will mimic the predator that's trying to eat them, so that the predator misinterprets them as a member of their own species and gives them territorial display instead of eating them.

They will even, I have to tell you, mimic the feces, or droppings, of their predators. That's so common it has a technical term in the literature, forgive me, "shit mimics." And they come in all varieties and sizes. There are moths that look like the splash variety of a bird dropping. And you can understand from the bird's standpoint, you might have a strong supposition that this is a butterfly or a moth, but you'd be unwilling to put it to the test—especially if you have to use your beak to put it to a test.

Now when you turn to relations within species, you find a rich world that we're uncovering now of deception also. To give you two quick examples: Warning cries have evolved in many contexts to warn others of danger. But they can be used in new and deceitful contexts. For example, you can give a warning cry in order to grab an item of food from another individual. The individual is startled and runs for cover, you grab the food. You can give a warning cry when your off-spring are at each other's throats—they run to cover and then you separate them and protect them from each other. It has even been described that you can give a warning call when you see your mate near a prospective lover—get them dashing to safety, and then you intervene.

In this continually co-evolving struggle regarding truth and falsehood, if you will, there are situations in other creatures as well as ourselves where we have to make tight evaluations of each others' motive in an aggressive encounter. I'm lining up against Marc Hauser; how confident is he of himself? I'm courting someone; the woman is looking at me; how confident am I of myself? And so on. That allows misrepresentation of these kinds of psychological variables and you can see how self-deception can start coming in. Be more confident than you have grounds to be confident, and be unconscious of that bias, the better to manipulate others.

Once you have language, that greatly increases the opportunity for both deception and self-deception. We spend a lot of time with each other pushing various theories of reality, which are often biased towards our own interests but sold as being generally useful and true.

Let me just mention a little bit of evidence—and of course there's a huge amount of evidence regarding self-deception, from everyday life, from study of politics and history, autobiography, et cetera. But I just want to talk about some of the scientific evidence in psychology. There's a whole branch of social psychology that's devoted to our tendencies for self-inflation. If you ask students how many of them think they're in the top half of the class in terms of leadership ability, eighty percent say they are. But if you turn to their professors and ask them how many think they're in the top half of their profession, ninety-four percent say they are.

And people are often unconscious of some of the mechanisms that naturally occur in them in a biased way. For example, if I do something that is beneficial to you or to others, I will use the active voice: I did this, I did that, then benefits rained down on you. But if I did something that harmed others, I unconsciously switch to a passive voice: this happened, then that happened, then unfortunately you suffered these costs. One example I always loved was a man in San Francisco who ran into a telephone pole with his car, and he described it to the police as, "the pole was approaching my car, I attempted to swerve out of the way, when it struck me."

Let me give you another, the way in which group membership can entrain language usages that are self-deceptive. You can divide people into in-groups or out-groups, or use naturally occurring in-groups and out-groups, and if someone is a member of your in-group and they do something nice, you give a general description of it: "he's a generous person." If they do something negative, you state a particular fact: "in this case he misled me," or something like that. But it's exactly the other way around for an out-group member. If an out-group member does something nice, you give a specific description of it: "she gave me directions to where I wanted to go." But if she does something negative, you say: "she's a selfish person." So these kinds of manipulations of reality are occurring largely unconsciously, in a way that's perhaps similar to what Marc Hauser in his talk was saying about morality.

A new world of the neurophysiology of deceit and self-deception is emerging. For example, it has been shown that consciously directed forgetting can produce results a month later and they are achieved by a particular area of the pre-frontal cortex, (normally associated with initiating motor responses or overcoming cognitive obstacles) suppressing activity in the hippocampus, the brain region in which memories are stored. So there is clear evidence that one part of the brain has been co-opted in evolution to serve the function of personal information suppression within self.

What I want to turn to very briefly is the relationship between self-deception and war. Now war, in the sense of battles between large numbers of soldiers, is an evolution-arily very recent phenomenon. A raid, where you run over to another group, kill off a number of individuals, and run back, is something we share with chimpanzees. And that has a long history and is much more likely to be constrained by rational considerations.

But warfare as we experience it now is a ten thousand, (plus or minus a few thousand) year old phenomenon. Not an awful lot of time for selection. And not much selection necessarily on those who start the wars. There may be a lot of selection in the civilian population or the soldiers, but it's not necessarily true that those who start stupid wars end up with as great a decrease in surviving offspring (and other kin) as one would have wished.

Wars also tempt us easily to self-deception for other reasons. There is often very little overlap in self-interest between your group and another group, in contrast to activities within the group. There is also low feedback from members of an outside group. There's greater ignorance.

And so war is a particular situation where self-deception is expected to be both especially prominent and especially harmful in its general effects.

Let us use the most recent war—the current war launched by my own country, the United States, in 2003 against the country of Iraq—to see one simple illustration of how deceit and self-deception is a useful concept in thinking about war. It has been said that the first casualty of war is the truth, but we know regarding the Iraq war that the truth was dead long before this war started. We know the thing was conceived and promulgated based on a lie. The predator, the U.S., saw an opportunity to leap on a prey, and decided almost immediately, within days of 9/11, and certainly within a couple of months, to prepare and launch this war.

Now what's the significance of that fact? Well, one significance of it is, psychologists have shown, very nicely I think, for twenty years now, that when we are considering an option—whether to marry Susan, or to go to the University of Bologna instead of Barcelona, or whatnot— we are much more rational, we weigh options, and we are even, if anything, slightly depressed. But once we decide which way to go, we act as if we want all the cells in our body rowing in the same direction. If it's Susan we're going to marry, we don't want to hear about Maria or some of Susan's less desirable sides. If it's Barcelona we're going to, that's the best university to go to and to hell with Bologna.

Now the point about this war is that there was no period of rational discussion of the pluses and minuses. The United States decided—at least a small cabal within it, including the President, decided—to go to war almost instantaneously. They immediately went into the implementation stage—your mood goes up, you downplay the negatives, after all, you have made your decision, and you do not wish to hear contrary opinions. You especially do not wish to hear contrary opinions if the real reasons for going to war can not be revealed and the whole public pretense is a lie.

Thus, all planning for the aftermath was dismissed because it greatly increased the apparent expense and difficulty and suggested greatly diminished gains from the endeavor. This, of course, implicitly called into question the entire enterprise, so rational planning was dismissed. And witness the dread effects, a continuing bloodbath unleashed on an innocent population.

One other comment: self-deception cannot only get you into disastrous situations, but then it gives you a second reward—and that is it deprives you of the ability to deal with the disaster once it's in front of you. And what could be more dramatic than what happened in the first month after the U.S. arrived in Baghdad—the complete looting of the country, twenty billion dollars of resources destroyed, priceless cultural heritage destroyed—all of that and the U.S. sat around and sucked its thumb. Did nothing to deal with it. And has been dealing with an escalating disaster ever since. A blood-letting of dreadful proportions, and still blind about what to do.

Well, I'll just summarize these thoughts by saying that there's good news and there's bad news. The good news is,

we do have it in our grasp at last to develop a scientific
theory of deceit and self-deception, integrating all kinds of
information, but at least sticking this phenomenon out in
front of ourselves and studying it objectively. The bad new is
that the forces we're dealing with—that is, of deceit and self-
deception—are very powerful.

PART II
MACHINA SAPIENS?

REBOOTING CIVILIZATION II

WITH SETH LLOYD, PAUL STEINHARDT, ALAN GUTH, MARVIN MINSKY, AND RAY KURZWEIL

In July of 2002, Edge *held an event at Eastover Farm, which included a lot of "universes" floating around: the computational universe, the cyclic universe, the inflationary universe, the emotional universe, and the intelligent universe. All, to some degree, were concerned with information processing and computation as central metaphors.*

Concepts of information and computation have infiltrated a wide range of sciences, from physics and cosmology to cognitive psychology and evolutionary biology. Such innovations as the binary code, the bit, and the algorithm have been applied in ways that reach far beyond the programming of computers and are being used to understand such mysteries as the origins of the universe, the operation of the human body, and the working of the mind.

Indeed, exactly what computation and information are continues to be subjects of intense debate. But less than a year later, in the "Week in Review" section of the Sunday *New York Times* ("What's So New in a Newfangled Science?" June 16, 2002), George Johnson wrote about "a movement some call digital physics or digital philosophy—a worldview that has been slowly developing for 20 years." The article continued:

> Just last week, a professor at the Massachusetts Institute of Technology named Seth Lloyd published a paper in *Physical Review Letters* estimating how many calculations the universe could have performed since the Big Bang—10^{120} operations on 10^{90} bits of data, putting the mightiest supercomputer to shame. This grand computation essentially consists of subatomic particles ricocheting off one another and "calculating" where to go.
>
> As the researcher Tommaso Toffoli mused back in 1984, "In a sense, nature has been continually computing the 'next state' of the universe for billions of years; all we have to do—and, actually, all we can do—is 'hitch a ride' on this huge ongoing computation."

This may seem like an odd way to think about cosmology. But some scientists find it no weirder than imagining that particles dutifully obey ethereal equations expressing the laws of physics. Last year Dr. Lloyd created a stir on Edge.org, a Web site devoted to discussions of cutting-edge science, when he proposed "Lloyd's hypothesis": "Everything that's worth understanding about a complex system can be understood in terms of how it processes information."

Dr. Lloyd did indeed cause a stir when his ideas were presented on *Edge* in 2001, but George Johnson's recent *New York Times* piece caused an even greater stir, as *Edge* received over half a million unique visits the following week, a strong confirmation that something is indeed happening here. (Usual *Edge* readership is about sixty thousand unique visitors a month). There is no longer any doubt that the metaphors of information processing and computation are at the center of today's intellectual action. A new and unified language of science is beginning to emerge.

THE COMPUTATIONAL UNIVERSE: SETH LLOYD

SETH LLOYD is professor of mechanical engineering at MIT and a principal investigator at the Research Laboratory of Electronics. He is also adjunct assistant professor at the Santa Fe Institute. He works on problems having to do with information and complex systems from the very small (how do atoms process information, how can you make them compute?) to the very large (how does society process information?).

His seminal work in the fields of quantum computation and quantum communications—including proposing the first technologically feasible design for a quantum computer, demonstrating the viability of quantum analog computation, proving quantum analogs of Shannon's noisy channel theorem, and designing novel methods for quantum error correction and noise reduction—has gained him a reputation as an innovator and leader in the field of quantum computing. Lloyd has been featured widely in the mainstream media, including the front page of the *New York Times*, the *LA Times*, the *Washington Post*, the *Economist*, *Wired*, the *Dallas Morning News*, and the *Times* (London). His name also frequently appears (both as writer and subject) in the *Nature*, *New Scientist*, *Science*, and *Scientific American*. He is the author of *Programming the Universe*: *A Quantum Computer Scientist Takes On the Cosmos*.

I build quantum computers that store information on individual atoms and then massage the normal interactions between the atoms to make them compute. Rather than having the atoms do what they normally do, you make them do elementary logical operations like bit flips, not operations, and-gates, and or-gates. This allows you to process information not only on a small scale but in ways that are not possible using ordinary computers. In order to figure out how to make atoms compute, you have to learn how to speak their language and to understand how they process information under normal circumstances. Every physical system registers information, and just by evolving in time, it changes that information, transforms that information, or, if you like, processes that information. Since I've been building quantum computers I've come around to thinking about the world in terms of how it processes information.

Say that you're building a computer out of some collection of atoms. How many logical operations per second could you perform? Also, how much information could these systems register? Using relatively straightforward techniques you can show, for instance, that the number of elementary logical operations per second that you can perform with that amount of energy, E, is just E/H—well, it's 2E divided by pi times h-bar (h-bar is essentially 10^{-34} [10 to the −34] joule-seconds, meaning that you can perform 10^{-50} ops per second). If you have a kilogram of matter, which has mc2—or around 10^{17}—joules worth of energy, and you ask how many ops per second it could perform, it could perform 10^{17} joules/h-bar. Using all the conventional techniques that were developed by Maxwell, Boltzmann, and Gibbs, and then developed by von Neumann and others back at the early part of the twentieth century for counting numbers of states, you can count how many bits it could register. What you find is that if you were to turn the thing into a nuclear fireball—which is essentially turning it all into radiation, probably the best way of having as many bits as possible—then you could register about 10^{30} bits. Actually that's many more bits than you could register if you just stored a bit on every atom, because Avogadro's number of atoms store about 10^{24} bits. I thought that it would be worthwhile to calculate how much information you could process if you were to use all the energy and matter of the universe.

I did this calculation, which was relatively simple. You take, first of all, the observed density of matter in the universe, which is roughly one hydrogen atom per cubic meter. The universe is about 13 billion years old, and using the fact that there are pi times 10^7 seconds in a year, you can calculate the total energy that's available in the whole universe. Remembering that there's a certain amount of energy, you then divide by Planck's constant—which tells you how many ops per second can be performed—and multiply by the age of the universe, and you get the total number of elementary logical operations that could have been performed since the universe began. You get a number that's around 10^{120}. It's a little bigger—10^{122}—but within astrophysical units, where if you're within a factor of one hundred, you feel that you're okay.

The other way you can calculate it is by calculating how it progresses as time goes on. The universe has evolved up to now, but how long could it go? One way to figure this out is to take the phenomenological observation of how much energy there is, but another is to assume, in a Guthian fashion, that the universe is at its critical density. Then there's a simple formula for the critical density of the universe in terms of its age; G, the gravitational constant; and the speed of light. If you plug this formula for the critical density into the expression for the number of ops that the universe can perform, then you find that the number of ops that could have been performed in the universe over time (T) since the universe began is actually the age of the universe divided by the Planck scale—the time at which quantum gravity becomes important—quantity squared. That is, it's the age of the universe squared, divided by the Planck length, quantity squared.

Having calculated the number of elementary logical operations that could have been performed since the universe began, I went and calculated the number of bits, which is a similar, standard sort of calculation. Say that we took all of this beautiful matter around us and vaporized it into a fireball of radiation. This would be the maximum entropy state and would enable the universe to store the largest possible amount of information. You can easily calculate how many bits could be stored by the amount of matter that we have in the universe right now, and the answer turns out to be 10^{90}. This is necessary, just by standard cosmological calculations—it's $(10^{120})^{3/4}$ (10 to the 120, quantity to the 3/4 power). We can store 10^{90} bits in matter, and if one believes in somewhat speculative theories about quantum gravity, such as holography—in which the amount of information that can be stored in a volume is bounded by the area of the volume divided by the Planck scale squared—then again you get 10^{120}. So the age of the universe squared, divided by the Planck time squared is equal to the size of the universe divided by the Planck length, quantity squared. So we can do 10^{120} (10 to the 120) ops on 10^{90} (10 to the 90) bits.

The third interpretation, which of course is more controversial, arises if we imagine that the universe is itself a computer and that what it's doing is performing a computation. If this is the case, these numbers say how big that computation is—how many ops have been performed on how many bits within the horizon since the universe began. Thinking of the universe as a computer is controversial. I don't see why it should be so controversial, because many books of science fiction have already regarded the universe as a computer. The universe is clearly not a computer with a Pentium inside. It's not an electronic computer, although of course it operates partly by quantum electrodynamics, and it's not running Windows—at least not yet. Some of us hope that never happens, if only because you don't want the universe as a whole to crash on you all of a sudden. Luckily, whatever operating system it has seems to be slightly more reliable so far. But if people try to download the wrong software, or upgrade it in some way, we could have some trouble.

So why is this controversial? For one, it seems to be making a statement that's obviously false. The universe is not an electronic digital computer, it's not running some operating system, and it's not running Windows. Why does it make sense to talk about the universe as performing a computation at all? There's one sense in which it's actually obvious that the universe is performing a computation. Take any physical system—say, a quarter, for example. The quarter can register a lot of information. It registers each atom in it, has a position that registers a certain amount of information, and can be heads or tails. Because the quarter is a physical system, it's also dynamic and evolves in time. It's easier to notice if I flip it in the air—it evolves in time, it changes, and as it changes it transforms that information, so the information that describes it goes from one state to another really fast. In addition, the positions, momentum, and quantum states of the atoms inside are changing, so the information that they're registering is changing. Merely by existing and evolving in time, any physical system registers information and transforms or processes that information.

It doesn't necessarily transform it or process it in the same way that a digital computer does, but it's certainly performing information processing. From my perspective, it's also uncontroversial that the universe registers 10^{90} bits of information, transforms and processes that information at a rate which is determined by its energy divided by Planck's constant. All physical systems can be thought of as registering and processing information, and how one wishes to define computation will determine your view of what computation consists of. If you think of computation as being merely information processing, then it's rather uncontroversial that the universe is computing, but of course many people regard computation as being more than information processing. There are formal definitions of what computation consists of. For instance, there are universal Turing machines, and there is a nice definition that's now seventy-odd years old of what it means for something to be able to perform digital computation. Indeed, the kind of computers we have sitting on our desks, as opposed to the kinds we have sitting in our heads or the kind that were in that little inchworm that was going along, are universal digital computers. So information processing, where a physical system is merely evolving in time, is a more specific, and potentially more powerful, kind of computing, because one way to evolve in time is just to sit there like a lump.

The other neat thing about these quantum computers is that they're also storing a bit of information on every available degree of freedom. Every nuclear spin in the molecules stores exactly one bit of information. We have examples of computers that saturate these ultimate limits of computation, and they look like actual physical systems. They look like alanine molecules, or amino acids, or like chloroform. Similarly, when we do quantum computation using photons, etc., we also perform computation at this limit.

I have not proved that the universe is, in fact, a digital computer and that it's capable of performing universal computation, but it's plausible that it is. It's also a reasonable scientific program to look at the dynamics of the standard model and to try to prove from that dynamics that it is computationally capable. We have strong evidence for this case. Why would this be interesting? It would also explain some things that have been otherwise paradoxical or confusing about the universe. Alan has done work for a long time on why the universe is so homogeneous, flat, and isotropic. This was unexplained within the standard cosmological model, and your great accomplishment here was to make a wonderful, simple, and elegant model that explains why the universe has these existing features. Another feature that everybody notices about the universe is that it's complex. Why is it complicated? Well, nobody knows. It turned out that way. Or if you're a creationist you say God made it that way. If you take a more Darwinian point of view the dynamics of the universe are such that as the universe evolved in time, complex systems arose out of the natural dynamics of the universe. So why would the universe being capable of computation explain why it's complex?

Computers are famous for being able to do complicated things starting from simple programs. You can write a very short computer program that will cause your computer to start spitting out the digits of pi. If you want to make it slightly more complex you can make it stop spitting out those digits at some point so you can use it for something else. There are short programs that generate all sorts of complicated things. That in itself doesn't constitute an explanation for why the universe itself exhibits all this complexity, but if you combine the fact that you have something that's dynamically, computationally universal with the fact that you're constantly having information injected into the universe—by the basic laws of quantum mechanics, quantum fluctuations are all the time injecting, programming the universe with bits of information—then you do have a reasonable explanation, which I'll close with.

About 120 years ago, Ludwig Boltzmann proposed an explanation for why the universe is complex. He said that it's just a big thermal fluctuation. His is a famous explanation: the monkeys-typing-on-typewriters explanation for the universe. Say there were a bunch of monkeys typing a bunch of random descriptions into a typewriter. Eventually we would get a book, right? But Boltzmann among other people realized right away that this couldn't be right, because the probability of this happening is vanishingly small. But now let's turn to this other metaphor: now the monkeys are not typing into a typewriter but into a computer keyboard. Let's suppose this computer is accepting what the monkeys are typing as instructions to perform computational tasks. This means that, for instance, because there are short programs for producing the digits of pi you don't need that many monkeys typing for that long until all of a sudden pi is being produced by the computer. There's a nice theory associated with this called algorithmic information theory, which says that if you've got monkeys typing into a computer the fact is that anything that can be realistically described by a mathematical equation, by a computer computing things, will at some point show up for these monkeys. In the monkey-typing-into-the-computer universe, all sorts of complex things arise naturally by the natural evolution of the universe.

I would suggest, merely as a metaphor here, but also as the basis for a scientific program to investigate the computational capacity of the universe, that this is also a reasonable explanation for why the universe is complex. It gets programmed by little random quantum fluctuations, like the same sorts of quantum fluctuations that mean that our galaxy is here rather than somewhere else. According to the standard model, billions of years ago some little quantum fluctuation, perhaps a slightly lower density of matter, maybe right where we're sitting right now, caused our galaxy to start collapsing around here. It was just a little quantum fluctuation, but it programmed the universe, and it's important for where we are, because I'm very glad to be here and not billions of miles away in outer space.

THE CYCLIC UNIVERSE:
PAUL STEINHARDT

PAUL STEINHARDT is the Albert Einstein Professor in Science and on the faculty of both the departments of Physics and Astrophysical Sciences at Princeton University.

He is one of the leading theorists responsible for inflationary theory. He constructed the first workable model of inflation and the theory of how inflation could produce seeds for galaxy formation. He was also among the first to show evidence for dark energy and cosmic acceleration, introducing the term *quintessence* to refer to dynamical forms of dark energy. Neil Turok pioneered mathematical and computational techniques that decisively disproved rival theories of structure formation, such as cosmic strings. He made leading contributions to inflationary theory and to our understanding of the origin of the matter-antimatter asymmetry in the universe. Hence, the authors not only witnessed but also led firsthand the revolutionary developments in the standard cosmological model caused by the fusion of particle physics and cosmology in the last twenty years.

He is the coauthor (with Neil Turok) of *Endless Universe: Beyond the Big Bang.*

If you were to ask most cosmologists to give a summary of where we stand right now in the field, they would tell you that we live in a very special period in human history where, thanks to a whole host of advances in technology, we can suddenly view the very distant and very early universe in ways that we haven't been able to do ever before. For example, we can get a snapshot of what the universe looked like in its infancy, when the first atoms were forming. We can get a snapshot of what the universe looked like in its adolescence, when the first stars and galaxies were forming. And we are now getting a full-detail, three-dimensional image of what the local universe looks like today. When you put together this different information, you obtain a very tight series of constraints on any model of cosmic evolution. If you go back to the different theories of cosmic evolution in the early 1990s, the data we've gathered in the last decade has eliminated all of them save one, a model that you might think of today as the consensus model. This model involves a combination of the Big Bang model as developed in the 1920s, '30s, and '40s; the inflationary theory, which Alan Guth proposed in the 1980s; and a recent amendment that I will discuss shortly. This consensus theory matches the observations we have of the universe today in exquisite detail. For this reason, many cosmologists conclude that we have finally determined the basic cosmic history of the universe.

But I have a rather different point of view, a view that has been stimulated by two events. The first is the recent amendment to which I referred earlier. I want to argue that the recent amendment is not simply an amendment, but a real shock to our whole notion of time and cosmic history. And secondly, in the last year I've been involved in the development of an alternative theory that turns the cosmic history topsy-turvy. All the events that created the important features of our universe occur in a different order, by different physics, at different times, over different time scales—and yet this model seems capable of reproducing all of the successful predictions of the consensus picture with the same exquisite detail.

The key difference between this picture and the consensus picture comes down to the nature of time. The standard model, or consensus model, assumes that time has a beginning that we normally refer to as the Big Bang. According to this model, for reasons we don't quite understand, the universe sprang from nothingness into somethingness, full of matter and energy, and has been expanding and cooling for the past 15 billion years. In the alternative model, the universe is endless. Time is endless in the sense that it goes on forever in the past and forever in the future, and, in some sense, space is endless. Indeed, our three spatial dimensions remain infinite throughout the evolution of the universe.

More specifically, this model proposes a universe in which the evolution of the universe is cyclic. That is to say, the universe goes through periods of evolution from hot to cold, from dense to under-dense, from hot radiation to the structure we see today, and eventually to an empty universe. Then, a sequence of events occurs that cause the cycle to begin again. The empty universe is reinjected with energy, creating a new period of expansion and cooling. This process repeats periodically forever. What we're witnessing now is simply the latest cycle.

During the attempts to try to bring cyclic ideas into modern cosmology, it was discovered in the '20s and '30s that there are various technical problems. The idea at that time was a cycle in which our three-dimensional universe goes through periods of expansion beginning from the Big Bang and then reversal to contraction and a big crunch. The universe bounces and expansion begins again. So, at the beginning of a new cycle, there is higher entropy density than the cycle before. It turns out that the duration of a cycle is sensitive to the entropy density. If the entropy increases, the duration of the cycle increases as well. So, going forward in time, each cycle becomes longer than the one before. The problem is that, extrapolating back in time, the cycles become shorter until, after a finite time, they shrink to zero duration. The problem of avoiding a beginning has not been solved. It has simply been pushed back a finite number of cycles. If we're going to reintroduce the idea of a truly cyclic universe, these two problems must be overcome. The cyclic model that I will describe uses new ideas to do just that.

To appreciate why an alternative model is worth pursuing, it's important to get a more detailed impression of what the consensus picture is like. Certainly some aspects are appealing. But, what I want to argue is that, overall, the consensus model is not so simple. In particular, recent observations have forced us to amend the consensus model and make it more complicated. So, let me begin with an overview of the consensus model.

The consensus theory begins with the Big Bang. It's a standard assumption that people have made over the last fifty years, but it's not something we can prove at present from any fundamental laws of physics. Furthermore, you have to assume that the universe began with an energy density less than the critical value. Otherwise, the universe would stop expanding and recollapse before the next stage of evolution, the inflationary epoch. In addition, to reach this inflationary stage, there must be some sort of energy to drive the inflation. Typically this is assumed to be due to an inflation field. You have to assume that in those patches of the universe that began at less than the critical density, a significant fraction of the energy is stored in inflation energy so that it can eventually overtake the universe and start the period of accelerated expansion. All of these are reasonable assumptions, but assumptions nevertheless.

Assuming these conditions are met, the inflation energy overtakes the matter and radiation after a few instants. The inflationary epoch commences and the expansion of the universe accelerates at a furious pace. The inflation does a number of miraculous things: it makes the universe homogeneous, it makes the universe flat, and it leaves behind certain inhomogeneities, which are supposed to be the seeds for the formation of galaxies. Now the universe is prepared to enter the next stage of evolution with the right conditions. According to the inflationary model, the inflation energy decays into a hot gas of matter and radiation. After a second or so, there form the first light nuclei. After a few tens of thousands of years, the slowly moving matter dominates the universe. It's during these stages that the first atoms form, the universe becomes transparent, and the structure in the universe begins to form—the first stars and galaxies. Up to this point, the story is relatively simple.

But, there is the recent discovery that we've entered a new stage in the evolution of the universe. After the stars and galaxies have formed, something strange has happened to cause the expansion of the universe to speed up again. During the 15 billion years when matter and radiation dominated the universe and structure was forming, the expansion of the universe was slowing down because the matter and radiation within it is gravitationally self-attractive and resists the expansion of the universe. Until very recently, it had been presumed that matter would continue to be the dominant form of energy in the universe, and this deceleration would continue forever.

But we've discovered instead, due to recent observations, that the expansion of the universe is speeding up. This means that most of the energy of the universe is neither matter nor radiation. Rather, another form of energy has overtaken the matter and radiation. For lack of a better term, this new energy form is called "dark energy." Dark energy, unlike the matter and radiation that we're familiar with, is gravitationally self-repulsive. That's why it causes the expansion to speed up rather than slow down.

I don't think either the physics or cosmology communities, or even the general public, has fully absorbed the full implications of this discovery. This is a revolution in the grand historic sense—in the Copernican sense. Copernicus's importance—from whom we derive the word *revolution*—was that he changed our notion of space and of our position in the universe. By showing that the earth revolves around the sun, he triggered a chain of ideas that led us to the notion that we live in no particular place in the universe; there's nothing special about where we are. Now we've discovered something very strange about the nature of time: that we may live in no special place, but we *do* live at a special time, a time of recent transition from deceleration to acceleration; from one in which matter and radiation dominate the universe to one in which they are rapidly becoming insignificant components; from one in which structure is forming in ever-larger scales to one in which now, because of this accelerated expansion, structure formation stops. We are in the midst of the transition between these two stages of evolution.

With these thoughts about the consensus model in mind, let me turn to the cyclic proposal. Since it's cyclic, I'm allowed to begin the discussion of the cycle at any point I choose. To make the discussion parallel, I'll begin at a point analogous to the Big Bang; I'll call it the Bang. This is a point in the cycle where the universe reaches its highest temperature and density. In this scenario, though, unlike the *Big*

Bang model, the temperature and density don't diverge. There is a maximal, *finite* temperature. It's a very high temperature, around 10^{20} (ten to the 20) degrees Kelvin—hot enough to evaporate atoms and nuclei into their fundamental constituents—but it's not infinite. The theory begins with a bang and then proceeds directly to a phase dominated by radiation. You still have to explain why the universe is flat, you still have to explain why the universe is homogeneous, and you still have to explain where the fluctuations came from that led to the formation of galaxies, but that's not going to be explained by an early stage of inflation.

In this new model, you go directly to a radiation-dominated universe and form the usual nuclear abundances; then go directly to a matter-dominated universe in which the atoms and galaxies and larger scale structure form; and then proceed to a phase of the universe dominated by dark energy. In the standard case, the dark energy comes as a surprise, since it is something you have to add into the theory to make it consistent with what we observe. In the cyclic model, the dark energy moves to center stage as the key ingredient that is going to drive the universe and in fact drives the universe into the cyclic evolution. The first thing the dark energy does when it dominates the universe is what we observe today: it causes the expansion of the universe to begin to accelerate. Over time the dark energy thins out the distribution of matter and radiation in the universe, making the universe more and more homogeneous and isotropic, driving it into what is essentially a vacuum state.

At the same time that the universe is made homogeneous and isotropic, it is also being made flat. If it continued forever, of course, that would be the end of the story. But in this scenario, just like inflation, the dark energy only survives for a finite period and triggers a series of events that eventually lead to a transformation of energy from gravity into new energy and radiation that will then start a new period of expansion of the universe.

Exactly how this works in detail can be described in various ways. I will choose to present a very nice geometrical picture that is motivated by superstring theory. We use only a few basic elements from superstring theory, so you don't really have to know anything about superstring theory to understand what I'm going to talk about, except to understand that some of the strange things that I'm going to introduce I am not introducing for the first time. They are already sitting there in superstring theory waiting to be put to good purpose.

One of the ideas in superstring theory is that there are extra dimensions; it's an essential element to that theory that is necessary to make it mathematically consistent. In one particular formulation of that theory the universe has a total of eleven dimensions. Six of them are curled up into a little ball so tiny that, for my purposes, I'm just going to pretend that they're not there. However, there are three spatial dimensions, one time dimension, and one additional dimension that I do want to consider. In this picture, our three dimensions with which we're familiar and through which we move lie along a hypersurface or membrane. This membrane is a boundary of the extra dimension. There is another boundary or membrane on the other side. In between, there's an extra dimension that, if you like, only exists over a certain interval. It's like we are one end of a sandwich, in between which there is a so-called bulk volume of space. These surfaces are referred to as *orbifolds* or *branes*—the latter referring to the word *membrane*. The branes have physical properties. They have energy and momentum, and when you excite them you can produce things like quarks and electrons.

We are composed of the quarks and electrons on these branes. And, since quarks and leptons can only move along branes, we are restricted to moving along and seeing only the three dimensions of our branes. We cannot see directly the bulk or any matter on the other brane.

In the cyclic universe, at regular intervals of trillions of years, these two branes smash together. This creates all kinds of excitations—particles and radiation. The collision thereby heats up the branes, and then they bounce apart again. The branes are attracted to each other through a force that acts just like a spring, causing the branes come together at regular intervals. At a microscopic distance away, there is another brane sitting and expanding, but since we can't touch, feel, or see across the bulk, we can't sense it directly. If there is a clump of matter over there, we can feel the gravitational effect, but we can't see any light or anything else that it emits, because anything it emits is going to move along that brane. We only see things that move along our own brane.

Next, the energy associated with the force between these branes takes over the universe. From our vantage point on one of the branes, this acts just like the dark energy we observe today. It causes the branes to accelerate in their stretching to the point where all the matter and radiation produced since the last collision is spread out, and the branes become essentially smooth, flat, empty surfaces. If you like, you can think of them as being wrinkled and full of matter up to this point and then stretching by a fantastic amount over the next trillion years. The stretching causes the mass and energy on the brane to thin out and the wrinkles to be smoothed out. After trillions of years, the branes are, for all intents and purposes, smooth, flat, parallel, and empty.

Then, the force between these two branes slowly brings the branes together. As it brings them together, the force grows stronger and the branes speed toward one another. When they collide, there's a walloping impact—enough to create a high density of matter and radiation with a very high, albeit finite, temperature. The two branes go flying apart, more or less back to where they are, and then the new matter and radiation (through the action of gravity) causes the branes to begin a new period of stretching.

In this picture it's clear that the universe is going through periods of expansion and a funny kind of contraction. Where the two branes come together, it's not a contraction of our dimensions but a contraction of the extra dimension. Before the contraction, all matter and radiation has been spread out, but, unlike the old cyclic models of the '20s and '30s, it doesn't come back together again during the contraction because our three dimensions—that is, the branes—remain stretched out. Only the extra dimension contracts. This process repeats itself cycle after cycle.

Remarkably, although the physical processes are completely different, and the time scale is completely different—this is taking billions of years, instead of 10^{-30} seconds—it turns out that the spectrum of fluctuations you get in the distribution of energy and temperature is essentially the same as what you get in inflation. Hence, the cyclic model is also in exquisite agreement with all of the measurements of the temperature and mass distribution of the universe that we have today.

Because the physics in these two models is quite different, there is an important distinction in what we would observe if one or the other were actually true—although this effect has not been detected yet. In inflation when you create fluctuations, you don't just create fluctuations in energy and temperature, but you also create fluctuations in space-time itself, so-called gravitational waves. In our model you don't get those gravitational waves. The essential difference is that inflationary fluctuations are created in a hyperrapid, violent process that is strong enough to create gravitational waves, whereas cyclic fluctuations are created in an ultraslow, gentle process that is too weak to produce gravitational waves. That's an example where the two models give an observational prediction that is dramatically different. It's just difficult to observe at the present time.

What's fascinating at the moment is that we have two paradigms that are now available to us. On the one hand they are poles apart, in terms of what they tell us about the nature of time, about our cosmic history, about the order in which events occur, and about the time scale on which they occur. On the other hand they are remarkably similar in terms of what they predict about the universe today. Ultimately what will decide between the two is a combination of observations—for example, the search for cosmic gravitational waves, and theory, because a key aspect to this scenario entails assumptions about what happens at the collision between branes that might be checked or refuted in superstring theory. In the meantime, for the next few years, we can all have great fun speculating about the implications of each of these ideas, which we prefer, and how we can best distinguish between them.

THE INFLATIONARY UNIVERSE: ALAN GUTH

ALAN GUTH, father of the inflationary theory of the universe, is the Victor F. Weisskopf Professor of Physics at MIT and author of *The Inflationary Universe: The Quest for a New Theory of Cosmic Origins.*

Paul Steinhardt did a very good job of presenting the case for the cyclic universe. I'm going to describe the conventional consensus model on which he was trying to say that the cyclic model is an improvement. I agree with what Paul said at the end of his talk as far as comparing these two models; it is yet to be seen which one works. But there are two grounds for comparing them. One is that in both cases the theory needs to be better developed. This is more true for the cyclic model, where one has the issue of what happens when branes collide. The cyclic theory could die when that problem finally gets solved definitively. Secondly, there is, of course, the observational comparison of the gravitational wave predictions of the two models.

Inflationary theory itself is a twist on the conventional Big Bang theory. The shortcoming that inflation is intended to fill in is the basic fact that although the Big Bang theory is called the Big Bang theory it is, in fact, not really a theory of a bang at all; it never was. The conventional Big Bang theory, without inflation, was really only a theory of the aftermath of the Bang. It started with all of the matter in the universe already in place, already undergoing rapid expansion, already incredibly hot. There was no explanation of how it got that way. Inflation is an attempt to answer that question, to say what "banged," and what drove the universe into this period of enormous expansion.

The basic idea behind inflation is that a repulsive form of gravity caused the universe to expand. General relativity predicts this repulsive form of gravity from the beginning: it's not just matter densities or energy densities that create gravitational fields; it's also pressures. A positive pressure creates a normal attractive gravitational field of the kind that we're accustomed to, but a negative pressure would create a repulsive kind of gravity. It also turns out that according to

modern particle theories, materials with a negative pressure
are easy to construct out of fields which exist according to
these theories. By putting together these two ideas—the fact
that particle physics gives us states with negative pressures,
and that general relativity tells us that those states cause
a gravitational repulsion—we reach the origin of the inflationary theory.

By answering the question of what drove the universe
into expansion, the inflationary theory can also answer some
questions about that expansion that would otherwise be very
mysterious. There are two very important properties of our
observed universe that were never really explained by the Big
Bang theory; they were just part of one's assumptions about
the initial conditions. One of them is the uniformity of the
universe—the fact that it looks the same everywhere, no
matter which way you look. It's both isotropic, meaning the
same in all directions, and homogeneous, meaning the same
in all places. To explain, for example, the uniformity of temperature that we see in the cosmic background radiation in
the standard Big Bang theory, you would need for that energy and information to transmit itself at about a hundred
times the speed of light just to allow the possibility that the
universe could have smoothed itself out and could have
achieved such a uniform temperature by the time the cosmic
background radiation was released.

In the inflationary theory this problem goes away completely, because in contrast to the conventional theory, it
postulates a period of accelerated expansion while this repulsive gravity is taking place. That means that if we follow our
universe backward in time toward the beginning using inflationary theory, we see that it started from something much
smaller than you ever could have imagined in the context of
conventional cosmology without inflation. While the region

that would evolve to become our universe was incredibly small, there was plenty of time for it to reach a uniform temperature, just like a cup of coffee sitting on the table cools down to room temperature. Once this uniformity is established on this tiny scale by normal thermo-equilibrium hypotheses—and I'm talking now about something that's about a billion times smaller than the size of a single proton—inflation can take over and cause that to expand rapidly and to become large enough to encompass the entire visible universe. The inflationary theory not only allows the possibility for the universe to be uniform, but also tells us why it's uniform: it's uniform because it came from something that had time to become uniform and was then stretched by this process of inflation.

The second peculiar feature of our universe that inflation does a wonderful job of explaining, and for which there never was a prior explanation, is the flatness of the universe and the fact that the geometry of the universe is so close to Euclidean. In the context of relativity, Euclidean geometry does not prevail; it's an oddity. With general relativity, curved space is the generic case. In the case of the universe as a whole, once we decide that the universe is homogeneous and isotropic, then this issue of flatness becomes directly related to the relationship between the mass density and the expansion rate of the universe. A large mass density would cause space to curve into a closed universe in the shape of a ball; if the mass density dominated, the universe would be a closed space with a finite volume and no edge. In the alternative case, if the expansion dominated, the universe would be geometrically open. Geometrically open spaces have the opposite geometric properties from closed spaces. They're infinite.

In terms of the evolution of the universe, the fact that the universe is at least approximately flat today requires that the early universe was extraordinarily flat. The universe tends to evolve away from flatness, so even given what we knew ten or twenty years ago—we know much better now that the universe is extraordinarily close to flat—we could have extrapolated backward and discovered that, for example, at one second after the Big Bang, the mass density of the universe must have been at the critical density where it counterbalanced the expansion rate to produce a flat universe. It must have been at that critical density to an accuracy of fifteen decimal places. The conventional Big Bang theory gave us no reason to believe that there was any mechanism to require that, and it has to have been that way to explain why the universe looks the way it does today.

There are two primary predictions that come out of inflationary models that are at least testable today. They have to do (1) with the mass density of the universe and (2) with the properties of these density non-uniformities. I'd like to say a few words about each of them, one at a time. Let me begin with the question of flatness.

The mechanism that inflation provides that drives the universe toward flatness will in almost all cases overshoot, not giving us a universe that is just nearly flat today but a universe that's almost exactly flat today. This can be avoided, and people have at times tried to design versions of inflation that avoided it, but to do so you have to go about it in a very contrived way. You have to arrange for inflation to end at just the right point, where it's almost made the universe flat but not quite. They always looked very contrived and never really caught on.

The generic inflationary model drives the universe to be completely flat, which means that one of the predictions is that today the mass density of the universe should be at the critical value which makes the universe geometrically flat.

Until three or four years ago no astronomers believed that. They told us that if you looked at just the visible matter, you would see only about 1 percent of what you needed to make the universe flat. But they also said that they could offer more than that—there's also dark matter. Dark matter is matter that's inferred to exist because of the gravitational effect that we see that it has on visible matter. You can ask how much mass is needed to hold those clusters of galaxies together, and the answer is that you still need significantly more matter than what you assumed was in the galaxies. Adding all of that together, astronomers came up only to about a third of the critical density. They were pretty well able to guarantee that there wasn't any more than that out there; that was all they could detect. That was bad for the inflationary model, but many of us still had faith that inflation had to be right and sooner or later the astronomers would come up with something.

And they did, although what they came up with was something very different from the kind of matter that we were talking about previously. Starting in 1998, the remarkable fact was observed that the universe today appears to be accelerating, not slowing down. As I said at the beginning of this talk, the theory of general relativity allows for that. What's needed is a material with a negative pressure. We're now therefore convinced that our universe must be permeated with a material with negative pressure, which is causing the acceleration that we're now seeing. We don't know what this material is, but we're referring to it as "dark energy." Even without knowing what it is, general relativity by itself allows us to calculate how much mass has to be out there to cause the observed acceleration, and it turns out to be almost exactly equal to two-thirds of the critical density. This is exactly what was missing from the previous calculations! Now, with the addition of the assumption that this dark energy is real, we have complete agreement between what the astronomers are telling us the mass density of the universe is and what inflation predicts.

The other important prediction that comes out of inflation is becoming even more persuasive than the issue of flatness: namely, the issue of density perturbations. Inflation has what in some ways is a wonderful characteristic—that by stretching everything out you can smooth out any non-uniformities that were present prior to this expansion. Inflation does not depend sensitively on what you assume existed before inflation; everything there just gets washed away by the enormous expansion. The universe really is fundamentally a quantum mechanical system, and it became clear that quantum theory was necessary not just to understand atoms but also to understand galaxies. It is rather remarkable that a fundamental idea like the basic ideas of quantum theory could have such a broad sweep. The point is that as inflation is ending, the classical Big Bang model would predict a completely uniform density of matter. According to quantum mechanics, however, everything is probabilistic. There are quantum fluctuations everywhere, which means that in some places the mass density would be slightly higher than average, and in other places it would be slightly lower than average. That's exactly the sort of thing you want to explain the structure of the universe. You can even go ahead and calculate the spectrum of these nonuniformities, which is something that Paul and I both worked on in the early days and had great fun with. The answer that we both came up with was that, in fact, quantum mechanics produces just the right spectrum of nonuniformities.

We really can't predict the overall amplitude—that is, the intensity of these ripples—unless we know more about the fundamental theory. At the present time, we have to take the overall factor that multiplies these ripples from observation. But we can predict the spectrum—that is, how the intensity of the ripples varies with the different wavelengths of all the different ripples that lie on top of each other. We knew how to do this back in 1982, but recently it has actually become possible for astronomers to see these nonuniformities imprinted on the cosmic background radiation. These were first observed back in 1992 by the Kobe satellite, but back then they could only see very broad features since the angular resolution of the satellite was only about seven degrees. Now, they've gotten down to angular resolutions of about a tenth of a degree. They basically plot the spectrum—the intensity of these ripples as a function of wavelength. Gradually these experimental plots are becoming more and more detailed.

The most recent data set was made by an experiment called the Cosmic Background Imager, which released a new set of data in May that is rather spectacular. This graph of the spectrum that I'm talking about is rather complicated because these fluctuations are produced during the inflationary era, but then oscillate as the early universe evolves.

At the present time, this inflationary theory, which a few years ago was in significant conflict with observation, now works perfectly with our measurements of the mass density and the fluctuations. The evidence for a theory that's either the one that I'm talking about or something very close to it is very, very strong.

I'd just like to close by saying that although I've been using the theory in the singular to talk about inflation, I shouldn't, really. It's very important to remember that inflation is really a class of theories. If inflation is right it's by no means the end of our study of the origin of the universe, but still, it's really closer to the beginning. There's still a lot of flexibility here, and a lot to be learned. And what needs to be learned will involve both the study of cosmology and the study of the underlying particle physics, which is essential to these models.

THE EMOTION UNIVERSE:
MARVIN MINSKY

MARVIN MINSKY, mathematician and computer scientist, is considered one of the fathers of artificial intelligence. He is Toshiba Professor of Media Arts and Sciences at the Massachusetts Institute of Technology; cofounder of MIT's Artificial Intelligence Laboratory; and the author of eight books, including *The Society of Mind* and *The Emotion Machine*.

I was listening to this group talking about universes, and it seems to me there's one possibility that's so simple that people don't discuss it. Certainly a question that occurs in all religions is, Who created the universe, and why? And what's it for? But something is wrong with such questions because they make extra hypotheses that don't make sense. So there's something wrong with questions like, What caused the universe to exist?

There's no need to think that our world "exists"; instead, think of it as like a computer game, and consider the following sequence of "Theories of It":

(1) Imagine that somewhere there is a computer that simulates a certain World, in which some simulated

people evolve. Eventually, when these become smart, one of those persons asks the others, "What caused this particular World to exist, and why are we in it?" But of course that World doesn't "really exist" because it is only a simulation.

(2) Then it might occur to one of those people that, perhaps, they are part of a simulation. Then that person might go on to ask, "Who wrote the Program that simulates us, and who made the Computer that runs that Program?"

(3) But then someone else could argue that, "Perhaps there is no Computer at all. Only the Program needs to exist because once that Program is written, then this will determine everything that will happen in that simulation. So the only real question is what is that program and who wrote it, and why?"

(4) Finally another one of those people observes, "No one needs to write it at all! It is just one of 'all possible computations'! So long as it is 'possible in principle,' then people in that Universe will think and believe that they exist!"

So we have to conclude that it doesn't make sense to ask about why this world exists. However, there still remain other good questions to ask, about how this particular universe works. For example, there cannot be structures that evolve (that is, in the Darwinian way) unless there can be some structures that can make mutated copies of themselves; this means that some things must be stable enough to have some persistent properties. Something like molecules that last long enough, etc.

So this, in turn, tells us something about physics: a universe that has people like us must obey some conservation-like laws; otherwise nothing would last long enough to support a process of evolution. We couldn't "exist" in a universe in which things are too frequently vanishing, blowing up, or being created in too many places. In other words, we couldn't exist in a universe that has the wrong kinds of laws.

THE CERTAINTY PRINCIPLE

In older times, when physicists tried to explain quantum theory to the public, what they call the uncertainty principle, they'd say that the world isn't the way Newton described it; instead they emphasized "uncertainty" that everything is probabilistic and indeterminate.

Yes, quantum theory shows that things are uncertain: if you have a DNA molecule there's a possibility that one of its carbon atoms will suddenly tunnel out and appear in Arcturus. However, at room temperature a molecule of DNA is almost certain to stay in its place for billions of years, because of quantum mechanics, and that is one of the reasons that evolution is possible! For quantum mechanics is the reason why most things don't usually jump around! Apparently, the first cells appeared quickly after the earth got cool enough; but then it took another 3 billion years to get to the kinds of cells that could evolve into animals and plants. This could only happen in possible worlds whose laws support stability. It could not happen in a Newtonian universe. So this is why the world that we're in needs something like quantum mechanics, to keep things in place!

INTELLIGENCE

Why don't we yet have good theories about what our minds are and how they work? In my view this is because we're only now beginning to have the concepts that we'll need for this.

Psychology itself did not much develop before the twentieth century. A few thinkers, like Aristotle, had good ideas about psychology, but progress thereafter was slow. Then came the era of Galton, Wundt, William James, and Freud, and we saw the first steps toward ideas about how minds work. But still, in my view, there was little more progress until the cybernetics of the '40s, the artificial intelligence of the '50s and '60s, and the cognitive psychology that started to grow in the '70s and '80s.

Perhaps psychology lagged behind because it tried to imitate the more successful sciences. For example, in the early twentieth century there were many attempts to make mathematical theories about psychological subjects, notably learning and pattern recognition. But there's a problem with mathematics. It works well for physics, I think, because fundamental physics has very few laws, and the kinds of mathematics that developed in the years before computers were good at describing systems based on just a few—say, four, five, or six laws—but doesn't work well for systems based on the order of a dozen laws. The beautiful subject called Theory of Groups begins with only five assumptions, yet this leads to systems so complex that people have spent their lifetimes on them. Similarly, you can write a computer program with just a few lines of code that no one can thoroughly understand; however, at least we can run the computer to see how it behaves, and sometimes see enough then to make a good theory.

However, there's more to computer science than that. Computer science is a new way to describe and think about complicated systems. It comes with a huge library of new, useful concepts about how mental processes might work.

Computer science suggests dozens of plausible ways to store knowledge away—as items in a database, sets of "if-then" reaction rules, in the forms of semantic networks (in which little fragments of information are connected by links that themselves have properties), program-like procedural scripts, or neural networks, etc. You can store things in what are called neural networks, which are wonderful for learning certain things but almost useless for other kinds of knowledge, because few higher-level processes can "reflect" on what's inside a neural network. This means that the rest of the brain cannot think and reason about what it's learned—that is, what was learned in that particular way. In artificial intelligence, we have learned many tricks that make programs faster but in the long run lead to limitations because the results of the neural network-type learning are too "opaque" for other programs to understand.

Yet even today, most brain scientists do not seem to know, for example, about cache-memory. If you buy a computer today you'll be told that it has a big memory on its slow hard disk, but it also has a much faster memory called cache, which remembers the last few things it did in case it needs them again, so it doesn't have to go and look somewhere else for them. And modern machines each use several such schemes, but I've not heard anyone talk about the hippocampus that way. All this suggests that brain scientists have been too conservative; they've not made enough hypotheses, and therefore most experiments have been trying to distinguish between wrong alternatives.

Reinforcement vs. Credit Assignment

There have been several projects that were aimed toward making some sort of "Baby Machine" that would learn and develop by itself to eventually become intelligent. However, all such projects, so far, have only progressed to a certain point and then became weaker or even deteriorated. One problem has been finding adequate ways to represent the knowledge that they were acquiring. Another problem was not having good schemes for what we sometimes call "credit assignment"—that is, how do you learn things that are relevant, that are essentials rather than accidents? For example, suppose that you find a new way to handle a screwdriver so that the screw remains in line and doesn't fall out. What is it that you learn? It certainly won't suffice merely to learn the exact sequence of motions (because the spatial relations will be different next time), so you have to learn at some higher level of representation. Clearly, one has to reinforce plans and not actions, which means that good credit assignment has to involve some thinking about the things that you've done. But still, no one has designed and debugged a good architecture for doing such things.

EXPERTISE VS. COMMON SENSE

In the early days of artificial intelligence, we wrote programs to do things that were very advanced. One of the first such programs was able to prove theorems in Euclidean geometry. This was easy because geometry depends only upon a few assumptions: Two points determine a unique line. If there are two lines then they are either parallel or they intersect in just one place. Or, two triangles are the same in all respects if the two sides and the angle between them are equivalent. This is a wonderful subject because you're in a world where assumptions are very simple, there are only a small number of them, and you use a logic that is very clear. It's a beautiful place, and you can discover wonderful things there.

However, I think that, in retrospect, it may have been a mistake to do so much work on tasks that were so "advanced." The result was that until today no one paid much attention to the kinds of problems that any child can solve. That geometry program did about as well as a superior high school student could do. Then one of our graduate students wrote a program that solved symbolic problems in integral calculus. Jim Slagle's program did this well enough to get a grade of A in MIT's first-year calculus course. (However, it could only solve symbolic problems, and not the kinds that were expressed in words.) However, in the mid 1960s, graduate student Daniel Bobrow wrote a program that could solve problems like "Bill's father's uncle is twice as old as Bill's father. Two years from now Bill's father will be three times as old as Bill. The sum of their ages is 92. Find Bill's age." Most high school students have considerable trouble with that. Bobrow's program was able to take convert those English sentences into linear equations and then solve those equations, but it could not do anything at all with sentences that had other kinds of meanings. We tried to improve that kind of program, but this did not lead to anything good because those programs did not know enough about how people use commonsense language.

Now, only in the past few years have a few researchers in AI started to work on the kinds of commonsense problems that every normal child can solve. But although there are perhaps a hundred thousand people writing expert specialized programs, I've found only about a dozen people in the world who aim toward finding ways to make programs deal with the kinds of everyday, commonsense jobs of the sort that almost every child can do.

THE INTELLIGENT UNIVERSE:
RAY KURZWEIL

RAY KURZWEIL was the principal developer of the first omni-font optical character recognition software, the first print-to-speech reading machine for the blind, the first CCD flat-bed scanner, the first text-to-speech synthesizer, the first music synthesizer capable of re-creating the grand piano and other orchestral instruments, and the first commercially marketed large vocabulary speech recognition. He has successfully founded, developed, and sold four AI businesses in OCR, music synthesis, speech recognition, and reading technology. All of these technologies continue today as market leaders.

Kurzweil received the $500,000 Lemelson-MIT Prize, the world's largest award in invention and innovation. He also received the 1999 National Medal of Technology, the nation's highest honor in technology, from President Clinton in a White House ceremony. He has also received scores of other national and international awards, including the 1994 Dickson Prize (Carnegie Mellon University's top science prize), Engineer of the Year from *Design News*, Inventor of the Year from MIT, and the Grace Murray Hopper Award from the Association for Computing Machinery. He has received ten honorary doctorate degrees and honors from three U.S. presidents. He has received seven national and international film awards. He is the author of *The Age of Intelligent Machines*; *The Age of Spiritual Machines, When Computers Exceed Human Intelligence*; and *The Singularity Is Near: When Humans Transcend Biology* and the coauthor (with Terry Grossman, M.D.) of *Fantastic Voyage: Live Long Enough to Live Forever.*

The universe has been set up in an exquisitely specific way so that evolution could produce the people that are sitting here today and we could use our intelligence to talk about the universe. We see a formidable power in the ability to use our minds and the tools we've created to gather evidence, to use our inferential abilities to develop theories, to test the theories, and to understand the universe at increasingly precise levels. That's one role of intelligence. The theories that we heard on cosmology look at the evidence that exists in the world today to make inferences about what existed in the past so that we can develop models of how we got here.

Then, of course, we can run those models and project what might happen in the future. Even if it's a little more difficult to test the future theories, we can at least deduce, or induce, that certain phenomena that we see today are evidence of times past, such as radiation from billions of years ago. We can't really test what will happen billions or trillions of years from now quite as directly, but this line of inquiry is legitimate, in terms of understanding the past and the derivation of the universe. As we heard today, the question of the origin of the universe is certainly not resolved. There are competing theories, and at several times we've had theories that have broken down, once we acquired more precise evidence.

At the same time, however, we don't hear discussion about the role of intelligence in the future. According to common wisdom, intelligence is irrelevant to cosmological thinking. It is just a bit of froth dancing in and out of the crevices of the universe and has no effect on our ultimate cosmological destiny. That's not my view. The universe has been set up exquisitely enough to have intelligence. There are intelligent entities like ourselves that can contemplate the universe and develop models about it, which is interesting. Intelligence is, in fact, a powerful force, and we can see that its power is going to grow not linearly but exponentially and will ultimately be powerful enough to change the destiny of the universe.

I want to propose a case that intelligence—specifically human intelligence, but not necessarily biological human intelligence—will trump cosmology, or at least trump the dumb forces of cosmology.

Commanding our local area of the sky is, of course, very small on a cosmological scale, but intelligence can overrule these physical forces, not by literally repealing the natural laws but by manipulating them in such a supremely sublime and subtle way that it effectively overrules these laws. This is particularly the case when you get machinery that can operate at nano and ultimately femto and pico scales. Whereas the laws of physics still apply, they're being manipulated now to create any outcome the intelligence of this civilization decides on.

Let me back up and talk about how intelligence came about. Wolfram's book has prompted a lot of talk recently on the computational substrate of the universe and on the universe as a computational entity. What Wolfram leaves out in talking about cellular automata is how you get intelligent entities. As you run these cellular automata, they create interesting pictures, but the interesting thing about cellular automata, which was shown long before Wolfram pointed it out, is that you can get apparently random behavior from deterministic processes.

It's more than apparent that you literally can't predict an outcome unless you can simulate the process. If the process under consideration is the whole universe, then presumably you can't simulate it unless you step outside the universe. But when Wolfram says that this explains the complexity we see in nature, it's leaving out one important step. As you run the

cellular automata, you don't see the growth in complexity—
at least, certainly he's never run them long enough to see any
growth in what I would call complexity. You need evolution.

Evolution works by indirection. It creates a capability and
then uses that capability to create the next. It took billions of
years until this chaotic swirl of mass and energy created the
information-processing, structural backbone of DNA and
then used that DNA to create the next stage. With DNA,
evolution had an information-processing machine to record
its experiments and conduct experiments in a more orderly
way. So the next stage, such as the Cambrian explosion, went
a lot faster, taking only a few tens of millions of years.

These designs worked well enough, so evolution could
then concentrate on higher cortical function, establishing
another level of mechanism in the organisms that could do
information processing. At this point, animals developed
brains and nervous systems that could process information,
and then that evolved and continued to accelerate. *Homo
sapiens* evolved in only hundreds of thousands of years, and
then the cutting edge of evolution again worked by indirection
to use this product of evolution, the first technology-creating
species to survive, to create the next stage: technology, a
continuation of biological evolution by other means.

The first stages of technologies, like stone tools, fire, and
the wheel took tens of thousands of years, but then we
had more powerful tools to create the next stage. A thousand
years ago, a paradigm shift like the printing press took only a
century or so to be adopted, and this evolution has accelerated ever since. Fifty years ago, the first computers were
designed with pencil on paper, with screwdrivers and wire.
Today we have computers to design computers. Computer
designers will design some high-level parameters, and twelve
levels of intermediate design are computed automatically.

The process of designing a computer now goes much more quickly.

Evolutionary processes accelerate, and the returns from an evolutionary process grow in power. I've called this theory the Law of Accelerating Returns. The returns, including economic returns, accelerate. Stemming from my interest in being an inventor, I've been developing mathematical models of this because I quickly realized that an invention has to make sense when the technology is finished, not when it was started, since the world is generally a different place three or four years later.

One exponential pattern that people are familiar with is Moore's Law, which is really just one specific paradigm of shrinking transistors on integrated circuits. It's remarkable how long it's lasted, but it wasn't the first, but the fifth paradigm to provide exponential growth to computing. Earlier, we had electromechanical calculators, using relays and vacuum tubes. Engineers were shrinking the vacuum tubes, making them smaller and smaller, until finally that paradigm ran out of steam because they couldn't keep the vacuum any more. Transistors were already in use in radios and other small, niche applications, but when the mainstream technology of computing finally ran out of steam, it switched to this other technology that was already waiting in the wings to provide ongoing exponential growth. It was a paradigm shift. Later, there was a shift to integrated circuits, and at some point, integrated circuits will run out of steam.

Moore's Law is one paradigm among many that have provided exponential growth in computation, but computation is not the only technology that has grown exponentially. We see something similar in any technology, particularly in ones that have any relationship to information. The genome project, for example, was not a mainstream project when it was announced. People thought it was ludicrous that you could scan the genome in fifteen years, because at the rate at which you could scan it when the project began, it could take thousands of years. But the scanning has doubled in speed every year, and actually most of the work was done in the last year of the project.

Another very important phenomenon is the rate of paradigm shift. This is harder to measure, but even though people can argue about some of the details and assumptions in these charts you still get these same very powerful trends. The paradigm shift rate itself is accelerating and roughly doubling every decade. When people claim that we won't see a particular development for a hundred years, or that something is going to take centuries to accomplish, they're ignoring the inherent acceleration of technical progress.

The last century is not a good guide to the next, in the sense that it made only about twenty years of progress at today's rate of progress, because we were speeding up to this point. At today's rate of progress, we'll make the same amount of progress as what occurred in the twentieth century in fourteen years, and then again in seven years. The twenty-first century will see, because of the explosive power of exponential growth, something like 20,000 years of progress at today's rate of progress—a thousand times greater than the twentieth century, which was no slouch for radical change.

Let's take computation. Communication is important and shrinkage is important. Right now, we're shrinking technology, apparently both mechanical and electronic, at a rate of 5.6 per linear dimension per decade. That number is also moving slowly, in a double exponential sense, but we'll get to nanotechnology at that rate in the 2020s. There are some early-adopter examples of nanotechnology today, but the real mainstream, where the cutting edge of the operating principles is in the multi-nanometer range, will be in the 2020s. If you put these together you get some interesting observations.

Right now we have 1,026 calculations per second in human civilization in our biological brains. We could argue about this figure, but it's basically, for all practical purposes, fixed. I don't know how much intelligence it adds if you include animals, but maybe you then get a little bit higher than 1,026. Nonbiological computation is growing at a double exponential rate and right now is millions of times less than the biological computation in human beings. Biological intelligence is fixed, because it's an old, mature paradigm, but the new paradigm of nonbiological computation and intelligence is growing exponentially. The crossover will be in the 2020s, and after that, at least from a hardware perspective, nonbiological computation will dominate at least quantitatively.

This brings up the question of software. Lots of people say that even though things are growing exponentially in terms of hardware, we've made no progress in software. But we are making progress in software, even if the doubling factor is much slower. The real scenario that I want to address is the reverse engineering of the human brain. Our knowledge of the human brain and the tools we have to observe and understand it are themselves growing exponentially. Brain scanning and mathematical models of neurons and neural structures are growing exponentially, and there's very interesting work going on.

It would be a mistake to say that the brain only has a few simple ideas and that once we can understand them we can build a very simple machine. But although there is a lot of complexity to the brain, it's also not vast complexity. It is described by a genome that doesn't have that much information in it. There are about 800 million bytes in the uncompressed genome. We need to consider redundancies in the DNA, as some sequences are repeated hundreds of thousands of times. By applying routine data compression, you can compress this information at a ratio of about 30 to 1, giving you about 23 million bytes—which is smaller than Microsoft Word—to describe the initial conditions of the brain.

But the brain has a lot more information than that. You can argue about the exact number, but I come up with thousands of trillions of bytes of information to characterize what's in a brain, which is millions of times greater than what is in the genome. How can that be? We know from computer science that we can very easily create programs of considerable complexity from a small starting condition. You can, with a very small program, create a genetic algorithm that simulates some simple evolutionary process and create something of far greater complexity than itself. You can use a random function within the program, which ultimately creates not just randomness but is creating some meaningful information after the initial random conditions are evolved using a self-organizing method, resulting in information that's far greater than the initial conditions.

The point of all of this is that, since it's a level of complexity we can manage, we will be able to reverse engineer the human brain. We've shown that we can model neurons, clusters of neurons, and even whole brain regions. We are well down that path. It's rather conservative to say that within twenty-five years we'll have all of the necessary scanning information and neuron models and will be able to put together a model of the principles of operation of how the human brain works. Then, of course, we'll have an entity that has some humanlike qualities. We'll have to educate and train it, but of course we can speed up that process, since we'll have access to everything that's out in the Web, which will contain all accessible human knowledge.

We can then combine some advantages of human intelligence with advantages that we see clearly in nonbiological intelligence. We spent years training our speech-recognition system, which gives us a combination of rules. It mixes expert-system approaches with some self-organizing techniques like neural nets, Markov models, and other self-organizing algorithms. We automate the training process by recording thousands of hours of speech and annotating it, and it automatically readjusts all its Markov-model levels and other parameters when it makes mistakes. Finally, after years of this process, it does a pretty good job of recognizing speech. Now, if you want your computer to do the same thing, you don't have to go through those years of training like we do with every child; you can actually load the evolved pattern of this one research computer, which is called loading the software.

The combination of the software of biological human intelligence with the benefits of nonbiological intelligence will be very formidable. Ultimately, this growing nonbiological intelligence will have the benefits of human levels of intelligence in terms of its software and our exponentially growing knowledge base.

In the future, maybe only one part of intelligence in a trillion will be biological, but it will be infused with human levels of intelligence, which will be able to amplify itself because of the powers of nonbiological intelligence to share its knowledge. How does it grow? Does it grow in or does it grow out? Growing in means using finer and finer granularities of matter and energy to do computation, while growing out means using more of the stuff in the universe. Presently, we see some of both. We see mostly the "in," since Moore's

Law inherently means that we're shrinking the size of transistors and integrated circuits, making them finer and finer. To some extent we're also expanding out in that even though the chips are more and more powerful, we make more chips every year, and deploy more economic and material resources towards this nonbiological intelligence.

Ultimately, we'll get to nanotechnology-based computation, which is at the molecular level, infused with the software of human intelligence and the expanding knowledge base of human civilization. It'll continue to expand both inward and outward. It goes in waves as the expansion inward reaches certain points of resistance. The paradigm shifts will be pretty smooth as we go from the second to the third dimension via molecular computing. At that point it'll be feasible to take the next step into femtoengineering (on the scale of trillionths of a meter) and pico engineering (on the scale of thousands of trillionths of a meter) going into the finer structures of matter and manipulating some of the really fine forces, such as strings and quarks. That's going to be a barrier, however, so the ongoing expansion of our intelligence is going to be propelled outward. Nonetheless, it will go both in and out. Ultimately, if you do the math, we will completely saturate our corner of the universe, the earth and solar system, sometime in the twenty-second century. We'll then want ever-greater horizons, as is the nature of intelligence and evolution, and will then expand to the rest of the universe.

How quickly will it expand? One premise is that it will expand at the speed of light, because that's the fastest speed at which information can travel. There are also tantalizing experiments on quantum disentanglement that show some effect at rates faster than the speed of light, even much faster, perhaps theoretically instantaneously. Interestingly enough, though, this is not the transmission of information but the transmission of profound quantum randomness, which doesn't accomplish our purpose of communicating intelligence.

If, in fact, that is a fundamental barrier, and if things that are far away really are far away, which is to say there are no shortcuts through wormholes through the universe, then the spread of our intelligence will be slow, governed by the speed of light. This process will be initiated within two hundred years. If you do the math, we will be at near saturation of the available matter and energy in and around our solar system, based on current understandings of the limitations of computation, within that time period. It may be very hard to do, but we're talking about supremely intelligent technologies and beings. If there are ways to get to parts of the universe through shortcuts such as wormholes, they'll find, deploy, and master them, and get to other parts of the universe faster. Then perhaps we can reach the whole universe, say 1,080 protons, photons, and other particles that Seth Lloyd estimates represents on the order of 1,090 bits, without being limited by the apparent speed of light.

If the speed of light is not a limit, and I do have to emphasize that this particular point is a conjecture at this time, then within three hundred years, we would saturate the whole universe with our intelligence, and the whole universe would become supremely intelligent and be able to manipulate everything according to its will. We're currently multiplying computational capacity by a factor of at least 103 every decade. This is conservative as this rate of exponential growth is itself growing exponentially. Thus it is conservative to project that within thirty decades (three hundred years), we would multiply current computational capacities by a factor of 1,090 and thus exceed Seth Lloyd's estimate of 1,090 bits in the universe. We can speculate about identity—will this be multiple people or beings, or one being, or will we all be merged?—but nonetheless, we'll be very intelligent and we'll be able to decide whether we want to continue expanding. Information is very sacred, which is why death is a tragedy. Whenever a person dies, you lose all that information in a person. The tragedy of losing historical artifacts is that we're losing information. We could realize that losing information is bad and decide not to do that anymore. Intelligence will have a profound effect on the cosmological destiny of the universe at that point.

Those that are ahead of us are not going to be ahead of us by only a few years. They're going to be ahead of us by billions of years. But because of the exponential nature of evolution, once we get a civilization that gets to our point, or even to the point of Babbage, who was messing around with mechanical linkages in a crude nineteenth-century technology, it's only a matter of a few centuries before they get to a full realization of nanotechnology, if not femto and pico engineering, and totally infuse their area of the cosmos with their intelligence. It only takes a few hundred years!

So if there are millions of civilizations that are millions or billions of years ahead of us, there would have to be millions that have passed this threshold and are doing what I've just said and have really infused their area of the cosmos. Yet we don't see them, nor do we have the slightest indication of their existence, a challenge known as the Fermi paradox. Someone could say that this "silence of the cosmos" is because the speed of light is a limit; therefore we don't see them, because even though they're fantastically intelligent, they're outside of our light sphere.

You might ask, isn't it incredibly unlikely that this planet, which is in a very random place in the universe and one of trillions of planets and solar systems, is ahead of the rest of the universe in the evolution of intelligence? Of course the whole existence of our universe, with the laws of physics so sublimely precise to allow this type of evolution to occur, is also very unlikely, but by the anthropic principles, we're here, and by an analogous anthropic principle, we are here in the lead. After all, if this were not the case, we wouldn't be having this conversation. So by a similar anthropic principle, we're able to appreciate this argument. I'll end on that note.

SOFTWARE IS A
CULTURAL SOLVENT

Jordan B. Pollack

I work on developing an understanding of biological complexity and how we can create it, because the limits of software engineering have been clear now for two decades. The biggest programs anyone can build are about 10 million lines of code. A real biological object—a creature, an ecosystem, a brain—is something with the same complexity as 10 billion lines of code. And how do we get there?

JORDAN B. POLLACK is a professor of computer science and complex systems at Brandeis University. His laboratory's work on AI, artificial life, neural networks, evolution, dynamical systems, games, robotics, machine learning, and educational technology has been reported on by the *New York Times*, *Time*, *Science*, NPR, and other media sources worldwide. Pollack is a prolific inventor, advises several start-up companies, and in his spare time runs Thinmail, which makes software to enhance e-mail and wireless telephone communications.

It's a marvelous age we live in—the age right before convergence with mechanism, when we will be wearing our computers as part of our bodies. People are talking now about Internet, television, and telephone conversion to personal wearable devices, but we are also within a century of the merger of bioinformatics, biotechnology, and information processing. As we understand cellular processes and neural representations and develop microelectronic and nanoscale technologies, our artifacts will be able to interact with our biology at a most fundamental level. Unfortunately, we haven't fathomed the complexity of nature yet well enough to know what to do with that.

I work on developing an understanding of biological complexity and how we can create it, because the limits of software engineering have been clear now for two decades. The biggest programs anyone can build are about 10 million lines of code. A real biological object—a creature, an ecosystem, a brain—is something with the same complexity as 10 *billion* lines of code. And how do we get there? My lab works on this question of self-organization, using evolution, neural networks, games, problem solving, and robotics. And the way we work is by trying to set up nonequilibrium chemical reactions in software which dissipate computer time—a form of energy—and create structure. Some of that structure we can make real, in the form of robots, and although robots are much more exciting to cameras and the media than problem solvers, games, and language learning are, our fundamental work is in trying to understand where complexity itself, without a designer, comes from.

The vision we're working on for robots is like that of the tool industry. There is no general-purpose tool; instead, there are drills, lathes, saws, routers, and other tools that fulfill specific purposes. We will make specific robots for specific goals. They won't be general-purpose Rosie Jetsons. They'll be something that might shovel your walk, clean your swimming pool, clean out the gutter, vacuum a room. There won't be any general-purpose, humanoid robots for centuries, in my opinion. Where we see things going—possibly in the next decade—is toward a robotic industry that will make hundreds of different dumb, special-purpose machines—things ultimately as sophisticated as ink-jet printers and ATMs, which are the real robots of today. My definition of a robot is a computer program hooked to a piece of hardware, working twenty-four hours a day and justifying the investment in its own creation. Robots may even put people out of work. In the case of ink-jet printers, the disemployed are scriveners; in the case of ATMs, they are bank tellers. So there is some disruption coming.

I've also been involved, since 1976, with microcomputers, the very small computers that are now everywhere; they're inside your cameras, they're inside tape recorders. Apple's new mouse has a supercomputer in it, apparently! We're coming to the age of wearable computers; I think the new devices—like the Blackberry and the PDA/phones—are really the beginning of wearable computers. You've seen people with cellphone buds in their ears all day long. These wearable computers are not what the pundits and nerds said that they would be, but people are carrying them around all day on their belts, using them all the time, and they will evolve into something approximating the science fiction communicator: voice with videos, MP3s, fax, and e-mail. It will be something you live with all day long, and we'll become untethered as a society.

Of course, to some people, wearable computers might sound like an electronic leash. But what's happening, inexorably, is that even though the world is supposedly smaller, our social-relationship networks are forming wider and wider nets. I have people in Atlanta, Washington, California, New York—all over the world, in fact—with whom I communicate, and I become more mobile and I travel more, and at the same time, the number of e-mail messages I process daily has increased from 10 to 100. And in a couple of years it's going to be 300 a day! We see a function for artificial intelligence here, as the e-mail message count and the cellphone message count rise.

For example, I've designed an adaptive e-mail filter. I can tell it that I want to see only 50 messages a day of those 300, but that I still want to know about the other 250. I don't want any machine to throw stuff away unless I've already said I don't want any more mail from this or that vendor. But I do want the top 50 in order of my priorities, and those are adaptive and change on a daily basis. I'll want to see a response, say, from someone I recently sent an e-mail to. The president of the United States might not be on my "in list," but I wouldn't want to miss an e-mail from him. Fairly simple AI techniques can essentially take those 300 messages a day and pick out the ones you're most likely to be interested in, while not throwing away anything you might have some chance of being interested in.

Demand for a wearable communication device that can intelligently deal with all your communication becomes more and more relevant. That's why the Blackberry is so popular now (it's the first popular wearable computer), since it almost seamlessly replicates the Microsoft Outlook view on your desktop in something that an executive can carry around. In a few years, we'll have eyepieces that give you a full-color view of your desktop on a little wearable wireless computer device. That's where it's going, and I'm excited to be involved in it.

Let's look at the problem of robotics from an economic point of view. Say that I could build you a vacuum-cleaner robot and it would cost $5 million to develop, and each vacuum would cost $5,000. But you can buy a plain old vacuum for $100 and push it yourself, or have someone who works for $8 an hour push it. There's no money to be made in robots; there's no mass market that will justify the kind of development necessary. Until you actually get a general-purpose humanoid robot that's cheap enough to hire instead of a human, there's a huge chasm to cross.

What my lab is doing is attempting to take the human engineers—the expensive, human, fixed-cost talent—out of the process of designing robots, so that we can make robots that are economical in a small quantity. A robot design may be needed only in one or three or five copies. A robot helping in a manufacturing production run may last only six months—not enough time to amortize the investment in its engineering. Only when we have robots that approach the cost of the materials needed to make them will they be economically valid without mass production. Our fully automatic design and manufacturing approach will not create the self-replicating robots of science fiction, but it uses software to design a machine in virtual reality to fulfill a particular purpose and then automatically builds that machine. This isn't scary; it's cost-effective. By doing automatic design, by having software that actually invents, we're beginning to raise the question that computer chess players raised a long time ago: If playing chess is really a human thing and a computer starts to do it, what does that tell us about humanity?

Life itself is an organizational principle and on a scale of complexity that dwarfs software engineering. We work on a basic biological question in my lab, which is, "How can a system dissipate energy and create more and more informational structure? How can a computer program write itself simply by wasting computer time?" In some sense, this is a computational, thermodynamic approach to artificial life, as opposed to a purely software-engineering approach to life. When we get to 10 billion lines of code, we will know whether we have succeeded or failed. The traditional notion—the high-church computation, the separation of the brain as hardware and the mind as software—is influential, but it's ultimately wrong, because the computational metaphor of serial programs operating on discrete data structures doesn't really capture the richness of natural systems. Our traditional Von Neumann notion is just not rich enough to capture what's going on in the natural world. I'm not saying that there's magical soul power there, just that we must drastically expand the notion of how information is represented beyond traditional symbolic computing.

We've been looking at chaotic systems, fractals, dynamical systems, both the attractors and the transients, all of which are very different from the traditional data-structure-plus-algorithm that you study in computer science. That idea is impoverished compared with what's happening in natural systems. What is God's cookbook that enables compositions of things to have new and surprising behaviors everywhere in the universe? Why do hydrogen and oxygen combine into something with a long liquid phase and such odd freezing properties? Random mixtures of organic chemicals are dense with behavioral possibilities, while random strings of machine code are 99.9 percent useless.

Traditional notions like the patent system are places where you can see this transition most effectively. A wheel is a piece of hardware: It's a thing, we've lathed it, it's round, it rolls, it carries a weight. But there's also a piece of software saying that "for every i from 1 to 360 plot r and theta, being the angle," and you've created a wheel in software! Now, traditionally the patent office has rejected algorithm as something that is of nature, something that is of God, and therefore unpatentable. But over the past decade and a half or so, we've allowed the patenting of software. We've even allowed the patenting of a business idea. Software is language, and we copyright it; it's language describing how a machine is going to work, all the way down to the lowest level of detail. And what a computer, a compiler, or an interpreter for programming languages does is to make our piece of text come alive; the machine will operate exactly the way the piece of text describes. A computer can take a description of a circle—the points that form a circle—and turn it into a virtual wheel inside a virtual environment. Software is a solvent, melting the boundary between what is virtual and what is real, between text and invention.

Some of the excitement over my lab's conception of automatically designed robots was the idea that we crossed from the virtual world back to the real. Software itself made inventions inside a computer—inventions that in another age could be patented. The end of that boundary between text and invention is something that profoundly affects society and the academy. Not only has software destroyed the text/invention distinction, software destroys the boundaries between what we used to own and what we used to rent. Property itself is being redefined in the information age. We used to buy a book and then we owned that book. But that book was really three different things fused into one. One part was the words, the information content; the second part was the medium, the physical delivery mechanism, the paper and the ink that captured the words; the third part was a social and legal contract, the license that said, "You may buy this book, and after you read it you may keep it in your library or pass it to a friend or sell it in a garage sale, but you can't make more copies to sell." Those three components— the medium, the content, and the license—have been torn asunder in the information age.

For instance, you might think you're buying a piece of software from Microsoft, but if you read the license carefully you'll see that it is a legal agreement saying that you didn't buy this, and what you did buy was a license to use it, and here are the conditions of use: You may not sell it separately from your computer. You may not make two copies of it. You can't give a copy to your friend while you're not using it, because it's part of your computer. And the actual content changes; it goes through forced upgrades, in which you have to buy it again and again, even though you've already bought it. And it comes on floppy disk, hard disk, Zip drive, compact flash; it comes downloaded off the Internet.

The fact that the three components of a book have been broken apart means that there are great opportunities for wealth (selling the same content over and over again) and also great opportunities for abuse. Both the Digital Rights Management and the File Sharing movements operate against property (as something you own until you sell it). I think the greater threat to human condition is not cheaper robots but the end of property, when our books, records, videos, software can no longer be owned. We need to move to a deeper understanding of property in the information age, as a bundle of rights. The onset of the Star Trek replicator, whose ancestors can be seen in today's rapid prototyping and 3-D printing machines, will mean that just like books, CDs, and software, hard objects will be copied. One day, Ford won't be a motor company; it will be an intellectual property company that licenses you a complex design for an arrangement of matter. You won't own your 2030 T-Bird, you'll only have licensed the right to keep atoms in that configuration for three years.

In the short term—say, the next five years—things won't be much different. In computers, we'll see pretty much the same kind of laptops we see now, just an incremental increase of power with more ports built in. In cellphones, we'll see better-integrated PDAs and e-mail systems, and we might see some interesting wireless multimedia. But like telegraphy, the entire network will be held back by the slowest common denominator, and text e-mail will be the Morse code of this era. I don't think much of 3G (the third generation of wireless data, which promises broadband speed), because it's too expensive to be practical, which is why I've put effort into commercializing one of my inventions.

And robotics? The robotics industry, as it exists today, caters expensive machinery to industries whose production profits are high enough to justify such luxuries—like chips, drugs, and software packaging. I don't see much really changing before 2005 in robotics, but by 2010 I think our vision of automated design and manufacture can have some impact. With the right investment and patience, I can see clearly how to create a general-purpose robotics industry that can automatically design and manufacture simple machines for industry and entertainment—an inversion of the traditional idea of building a humanoid robot slave that can do everything. A technology that can very cheaply produce specific dumbots for different kinds of tasks—assembly tasks, military tasks, cleanup tasks, entertainment tasks, even domestic tasks—might actually lead to a profitable and self-sustaining industry and a change in culture back toward the invention and manufacturing of *real* goods, instead of the dot-com stuff.

THE SECOND COMING—
A MANIFESTO

DAVID GELERNTER

The theme of the Second Age, now approaching, is that computing transcends computers. Information will travel through a sea of anonymous, interchangeable computers like a breeze through tall grass. A desktop computer will be a scooped-out hole in the beach, where information from the cybersphere wells up like seawater.

DAVID GELERNTER, a professor of computer science at Yale University and chief scientist at Mirror Worlds Technologies, is a leading figure in the third generation of artificial-intelligence researchers and the inventor of a programming language called Linda, which made it possible to link computers to work on a single problem. He has since emerged as one of the seminal thinkers in the field known as parallel, or distributed, computing. His books include *Mirror Worlds*; *The Muse in the Machine*; *Drawing a Life*; and *1939: The Lost World of the Fair.*

Any microsecond now, computing will be transformed. It's not just that our problems are big; they are big and *obvious*. It's not just that the solutions are simple; they are simple and right under our noses. And it's not that hardware is more advanced than software: The last big operating systems breakthrough was the Macintosh almost twenty years ago; today's hottest item is Linux—a version of Unix, which was new in 1976. Commercial software applications tend to be badly designed, badly made, incomprehensible, and obsolete. And users react to that hard truth by blaming themselves (*Computers for Morons, Operating Systems for Livestock*). But change is coming, and soon.

No matter how certain its eventual coming, an event whose exact time and form of arrival are unknown vanishes when we picture the future. We tend not to believe in the next big war, the next big economic swing; we certainly don't believe in the next big software revolution. Because we don't believe in technological change (we only say we do), we accept bad computer products with a shrug. We work around them, make the best of them, and (like fatalistic sixteenth-century French peasants) barely even notice their defects, instead of demanding that they be fixed and changed.

Everything is up for grabs. Everything will change. The Orwell law of the future: Any new technology that *can* be tried *will* be. Like Adam Smith's invisible hand leading capitalist economies toward ever-increasing wealth, Orwell's Law is a fact of life. Miniaturization was the big theme in the First Age of computers: rising power, falling prices, computers for everybody. The theme of the Second Age, now approaching, is computing transcends computers. Information will travel through a sea of anonymous, interchangeable computers like a breeze through tall grass. A desktop computer will be a scooped-out hole in the beach, where in formation from the cybersphere wells up like seawater. We'll become less and less interested in computers. The real topic in astronomy is the cosmos, not telescopes. The real topic in computing is the cybersphere and the cyberstructures in it, not the computers we use as telescopes and tuners.

The software systems we depend on most today are operating systems (Linux, Macintosh OS, Windows, et al.) and browsers (Internet Explorer, Netscape Communicator, et al.). Operating systems are connectors that fasten users to computers; they attach to the computer at one end, the user at the other. Browsers fasten users to remote computers, to "servers" on the Internet. Today's operating systems and browsers are obsolete because people no longer want to be connected to computers—nearby ones *or* remote ones. They probably never did. They want to be connected to information. The computing future is based on cyberbodies—self-contained, neatly ordered, beautifully laid out collections of information, like immaculate giant gardens. You will walk up to any "tuner" (a computer at home, work, or the supermarket; or a TV, a telephone, any kind of electronic device) and slip in a "calling card," which identifies a cyberbody. The tuner tunes it in. The cyberbody arrives and settles in, like a bluebird perching on a branch.

Your whole electronic life will be stored in a cyberbody. You will be able to summon it to any tuner at any time. By slipping in your calling card, you will customize any electronic device you touch; for as long as it holds your card, the machine will know your habits and preferences better than you know them yourself. The future will be dense with computers. They will hang around everywhere in lush growths, like Spanish moss. They will swarm like locusts. But a swarm is not merely a big crowd: Individuals in the swarm lose their identities; the computers that make up this global swarm will blend together into the seamless substance of the cybersphere. Within the swarm, individual computers will be as anonymous as molecules of air. A cyberbody will be replicated or distributed over many computers; it can inhabit many computers at the same time. If the cybersphere's computers are tiles in a paved courtyard, a cyberbody is a cloud's shadow drifting across many tiles simultaneously.

The Internet will change radically before it dies. When you deal with a remote Web site, you largely bypass the power of your desktop in favor of the far-off power of a Web server. Using your powerful desktop computer as a mere channel to reach Web sites—reaching through and beyond it, instead of using it—is like renting a Hyundai and keeping your Porsche in the garage, like executing programs out of disk storage instead of main memory and cache. The Web makes the desktop impotent. But the power of desktop machines is a magnet that will reverse today's "Everything onto the Web!" trend. Desktop power will inevitably drag information out of remote servers onto desktops. If a million people use a Web site simultaneously, does that mean we need a heavy-duty remote server to keep them all happy? No. We could move the site onto a million desktops and use the Internet for coordination. The site is like a military unit in the field, the general moving with his troops, or like a hockey team in constant swarming motion. (We used essentially this technique to build the first tuple space implementations. They seemed to depend on a shared server, but the server was an illusion; there was no server, just a swarm of clients.) Could Amazon.com be an itinerant horde instead of a fixed central command post? Yes.

PROBLEMS ON THE SURFACE AND UNDER THE SURFACE

The windows/menus/mouse/desktop interface, invented by Xerox and Apple and now universal, was a brilliant invention and is now obsolete. It wastes screen space on meaningless images, fails to provide adequate clues to what is inside the files represented by those blurry little images, forces users to choose icons for the desktop when the system could choose them better itself, and keeps users jockeying windows like parking attendants rearranging cars in a pint-sized Manhattan lot, in a losing battle for an unimpeded view of the workspace, which is ultimately unattainable. No such unimpeded view exists.

Icons and "collapsed views" seem new, but we have met them before. Any book has a "collapsed" or "iconified" view—namely, its spine. An icon conveys far less information than the average book spine and is much smaller. *Should* it be much smaller? Might a horizontal stack of "book spines" on-screen be more useful than a clutter of icons?

The computer mouse was a brilliant invention, but we can see today that it is a bad design. Like any device that must be moved and placed precisely, it ought to provide tactile feedback; it doesn't.

Metaphors have a profound effect on computing. The desktop metaphor traps us in a broad instead of a deep arrangement of information, which is fundamentally wrong for computer screens. A desktop is easily extended (you use drawers, other desks, tables, the floor); a computer screen is not. Apple could have described its interface as an "information landscape" instead of a "desktop." We invented this landscape (they might have explained) the way a landscape architect or theme park designer invents a landscape. We invented an ideal space for seeing and managing computerized information. Our landscape is imaginary, but you can still enter it and move around in it. The computer screen is the window of your vehicle, the face shield of your diving helmet. But with the desktop metaphor, the screen *is* the interface—a square foot or two of glowing colors on a glass panel. In the landscape metaphor, the screen is just a viewing pane. When you look through it, you see the actual interface lying beyond.

Modern computing is based on an analogy between computers and file cabinets, which is wrong and affects nearly every move we make. Computers are fundamentally unlike file cabinets, because they can *act*. They are machines, not furniture. The file-cabinet metaphor traps us in a passive instead of active mode of information management. The rigid file-and-directory system you are stuck with on your Mac or PC was designed by programmers for programmers—and is still a good system for programmers. It is no good for non-programmers. It never was, and was never intended to be.

If you have three pet dogs, give them names. If you have 10,000 head of cattle, don't bother. Nowadays the idea of giving a name to every file on your computer is ridiculous. The standard policy on file names has far-reaching consequences: It doesn't merely force us to make up names where no name is called for; it also imposes limits on our handling of an important class of documents—ones that arrive from the outside world. A newly arrived e-mail message (for example) can't stand on its own as a separate document—can't show up alongside other files in searches, sit by itself on the desktop, be opened or printed independently. It has no name, so it must be buried on arrival inside some existing file (the mail file) that does have a name. The same holds for incoming photos and faxes, Web bookmarks, scanned images, et cetera.

You shouldn't have to put files in directories. The directories should reach out and take them. If a file belongs in six directories, all six should reach out and grab it automatically, simultaneously. A file should be allowed to have no name, one name, or many names. Many files should be allowed to share one name. A file should be allowed to be in no directory, one directory, or many directories. Many files should be allowed to share one directory. Of these eight possibilities, only three are legal and the other five are banned—for no good reason.

LIFESTREAMS

In the beginning, computers dealt mainly in numbers and words. Today they deal mainly with pictures. In the period now emerging, they will deal mainly with tangible time—

time made visible and concrete. Chronologies and timelines tend to be awkward in the off-computer world of paper, but they are natural online.

Computers make alphabetical order obsolete. File cabinets and human minds are both information storage systems. We could model computerized information storage on the mind instead of the file cabinet, if we wanted to. Elements stored in a mind do not have names and are not organized into folders; they are retrieved not by name or folder but by contents. (Hear a voice, think of a face: You've retrieved a memory that contains the voice as one component.) You can see everything in your memory from the standpoint of past, present, and future. Using a file cabinet, you classify information when you put it in; minds classify information when it is taken out. (Yesterday afternoon at four, you stood with Natasha on Fifth Avenue in the rain—as you might recall when you are thinking about Fifth Avenue, rain, or Natasha. But you attached no such labels to the memory *when you acquired it.* The classification happened retrospectively.)

A "lifestream" organizes information—not as a file cabinet does but roughly as a mind does. A lifestream is a sequence of all kinds of documents—all the electronic documents, digital photos, applications, Web bookmarks, Rolodex cards, e-mail messages, and every other digital information chunk in your life—arranged from oldest to youngest, constantly growing as new documents arrive, easy to browse and search, and with a past, present, and future appearing on your screen as a receding parade of index cards. Documents have no names and there are no directories. You retrieve elements by content: "Fifth Avenue" yields a substream of every document that mentions Fifth Avenue. A substream (for example, the "Fifth Avenue" substream) is like a conventional directory, except that it builds itself automatically. It traps new documents as they arrive.

One document can be in many substreams, and a substream has the same structure as the main stream: a past, present, and future—steady flow.

A stream flows because time flows, and the stream is a concrete representation of time. A "Now" line divides past from future. If you have a meeting at 10:00 am tomorrow, you put a reminder document in the future of your stream, at 10:00 am tomorrow. It flows steadily toward "Now." When "Now" equals 10:00 am tomorrow, the reminder leaps over the "Now" line and flows into the past. When you look at the future of your stream, you see your plans and appointments, flowing steadily out of the future into the present, then the past. A lifestream is a landscape you can navigate or fly over at any level. Flying toward the start of the stream is time travel into the past.

You manage a lifestream by using two basic controls, "put" and "focus", which correspond roughly to acquiring a new memory and remembering an old one. To send e-mail, you put a document on someone else's stream. To add a note to your calendar, you put a document in the future of your own stream. To continue work on an old document, put a copy at the head of your stream. Sending e-mail, updating the calendar, opening a document are three instances of the same operation: putting a document in a stream.

The point of lifestreams isn't to shift from one software structure to another but to shift the whole premise of computerized information: to stop building glorified file cabinets and start building simplified, abstract artificial minds, in which to store our electronic lives. The lifestream (or some other system with the same properties) will become the most important information-organizing structure in computing, because even a rough imitation of the human mind is vastly more powerful than the most sophisticated file cabinet ever conceived. Lifestreams (in preliminary form) are a successful commercial product today, but my predictions have nothing to do with this product. Ultimately the product may succeed or fail; the idea will succeed. In late 2002, we released a beta version of our desktop lifestream system and saw 10,000 downloads in two weeks; it's no longer quite impossible to believe that our own software will be the winner.

Movies, TV shows, virtual museums, and all sorts of other cultural products—from symphonies to baseball games—will be stored in lifestreams. Institutions, too, will be afloat in the cybersphere. Your car, your school, your company, and yourself are all one-track vehicles moving forward through time, each leaving a stream-shaped cyber-body (like an aircraft's contrail) behind it as it goes. Those vapor trails of crystallized experience will represent our first concrete answer to a hard question: What *is* a company, a university, any sort of ongoing organization or institution, if its staff and customers and owners can all change, its build-ings can be bulldozed, its site can be relocated? What's left? What *is* it? The answer: a lifestream in cyberspace. Every employee has a private view of (and access to) the communal company stream. The company's Web site will be the pub-licly accessible substream of the main company stream. The company's lifestream is an electronic approximation of the company's memories, its communal mind.

Lifestreams won't yield the paperless office. The paper-less office is a bad idea, because paper is one of the most u seful and valuable media ever invented. But lifestreams will turn office paper into a *temporary* medium—for use, not storage. "On paper" is a good place for information you want to use; it's a bad place for information you want to store. In the stream-based office, you will scan each newly created or newly received paper document into the stream and throw the paper version away. When you need a paper document, you'll find it in the stream, print it out, use it. If you wrote on the paper while you were using it, you'll scan it back in and throw it out.

Nowadays we use a scanner to transfer a document's electronic image into a computer. Soon the scanner will become a cybersphere port of entry, an all-purpose in-box. Put any object in the in-box and the system will develop an accurate three-dimensional physical transcription and drop the transcription into the cool dark well of cyberspace. So the cybersphere starts to take on just a hint of the textural richness of real life. We'll know the new system is working when a butterfly wanders into the in-box and a few wingbeats later flutters out—and in that brief interval the system will have transcribed the creature's appearance and analyzed its way of moving. The real butterfly will leave a shadow butterfly behind. Some time soon afterward, you'll be examining some tedious electronic document and a cyberbutterfly will appear at the bottom left corner of your screen (maybe a *Hamearis lucina*) and pause there, briefly hiding the text and showing its neatly folded rusty-chocolate wings with orange eyespots, like a Victorian paisley—and moments later will have crossed the screen and vanished.

Imagine the future this way. If you have plenty of money, the best consequence (so they say) is that you no longer need to think about money. In the future, we will have plenty of technology—and the best consequence will be that we will no longer have to think about technology. We will return with gratitude and relief to the topics that actually count.

MAKING LIVING SYSTEMS

RODNEY BROOKS

My midlife research crisis has been to move away from looking at humanoid robots and toward looking at the very simple question of what makes something alive—what the organizing principles are that go on inside living systems. In my lab at MIT, we're trying to build robots that have properties of living systems that robots haven't had before.

RODNEY BROOKS is the director of the MIT Artificial Intelligence Laboratory and Fujitsu Professor of Computer Science at MIT. He is also chairman and chief technical officer of iRobot, a robotics company. Dr. Brooks appeared as one of the four principals in the 1997 Errol Morris movie *Fast, Cheap, and Out of Control* (named after one of Brooks's papers in the *Journal of the British Interplanetary Society*). He is the author of *Flesh and Machines* and *Cambrian Intelligence*.

Every nine years or so, I change what I'm doing scientifically. In 2001 I moved away from building humanoid robots to think about the difference between living matter and nonliving matter. Over here is an organization of molecules and it's a living cell; over there is an organization of molecules and it's just matter. What is it that makes something alive?

We have all become computationcentric over the last few years. We've tended to think that computation explains everything. When I was a kid, I had a book that described the brain as a telephone-switching network. Earlier books described it as a hydrodynamic system or a steam engine. Then in the 1960s it became a digital computer. In the 1980s it became a massively parallel digital computer. Probably there's a kids' book out there somewhere that says the brain is just like the World Wide Web because of all its associations. We're always taking the best technology we have and using that as the metaphor for the most complex thing we know—the brain. And now we're talking about computation.

But maybe there's more to us than computation. Maybe there's something beyond computation—in the sense that we do not understand and cannot describe what's going on inside living systems using computation only. When we build computational models of living systems—such as a self-evolving system or an artificial immunology system—they're not as robust or rich as real living systems. Maybe we're missing something, but what could that something be?

You could hypothesize that what's missing might be some aspect of physics that we don't yet understand. The philosopher David Chalmers has used that notion when he tries to explain consciousness. The mathematician Roger Penrose uses that notion to a certain extent when he says that thought arises from quantum effects in the microtubules of our neurons; he is looking for some physics we already understand but are just not describing well enough.

In the time of Kepler and Copernicus, people could describe what was happening in the solar system using observation and geometry and algebra, but it wasn't until they had calculus that they had a good model of what was happening and could make predictions. My working hypothesis is that in our understanding of complexity we're stuck at that algebra/geometry stage. There's some other tool—some organizational principle—that we need to understand in order to describe what's really going on.

So my midlife research crisis has been to move away from looking at humanoid robots and toward looking at the very simple question of what makes something alive—the organizing principles that go on inside living systems. In my lab at MIT, we're trying to build robots that have properties of living systems that robots haven't had before. It's an interdisciplinary lab, with students from across the institute, although the vast majority are computer science majors. We have electrical engineering majors, brain and cognitive science students, mechanical engineering students, even some aeronautics and astronautics students these days, because there's a push for autonomous systems in space. We work on a mixture of applied and theoretical stuff. We're trying to build robots that can repair themselves, that can reproduce (although we're a long way from self-reproduction), that metabolize, that go out and seek energy to maintain themselves. We have a robot now that wanders the corridors, finds electrical outlets, and plugs itself in. The next step is to make it hide during the day and come out at night and plug itself in. We're trying to build robots not out of silicon and steel but out of materials that are less rigid, less traditional—materials that are more like what we are built out of. Our mantra is that we're going to build a robot out of Jell-O. Not literally, but that's the image we have in mind.

Our work combines theory and application. A robot we're getting into production at iRobot, after three years of testing, goes down into oil wells. It's five centimeters in diameter and fourteen meters long, and it has to be autonomous, because you can't communicate down there by radio. As things are now, if you want to manipulate oil wells while they're in production, you need infrastructure on the surface to shove a thick cable down the well. This can mean miles and miles of cable, which means tons of cable on the surface, or a ship sitting above the oil well, pushing this stuff down through thirty-foot segments of pipe that go one after the other after the other for days and days. Our robots can go down oil wells—where the pressure is 10,000 psi at 150 degrees centigrade—and carry instruments, make various measurements, find out where there might be too much water coming into the well. When you have a managed oil well, you can increase production over the life of the well by about a factor of two, but it has been far too expensive to manage oil wells, because you need this massive infrastructure. These robots cost something on the order of $100,000. They're retrievable, because you don't want them down there blocking the oil flow. Even a robot five centimeters in diameter in an oil bore of standard size soon starts to clog things up. You can't communicate with them, but we've pushed them to failures artificially and also had some unpredicted failures down there, and in every case they've managed to reconfigure themselves and get themselves out.

The most successful applications over the past five years have been in surgery. Using computer-vision techniques, we have built robots that take all different sorts of imagery during surgery. You can have a patient inside an MRI as you're doing surgery. You get coarse measurements, register those with the fine MRI measurements done in a bigger machine beforehand, and then give the surgeon a real-time three-dimensional picture of everything inside the brain of a patient undergoing brain surgery. If you go to one of the major Boston hospitals for brain surgery, your surgeon will be assisted by AI systems developed in our lab. The first few times this equipment was running, there were graduate students in the operating room rebooting Unix at critical points. Now we're way past that, and none of our staff needs to be there anymore. It's all handed over to the surgeons and the hospital staff and it's working well.

We are also building large-scale computational experiments. People might call them simulations, but since we're not necessarily simulating anything real, I prefer to call them experiments. We're looking at a range of questions about living systems. One student, for example, is looking at how multicellular reproduction can arise from single-cell reproduction.

You can see how single-cell reproduction works, but how did that turn into multicellular reproduction, which at one level of organization looks very different from what's happening in single-cell reproduction? In single-cell reproduction, something gets bigger and then breaks in two; in multicellular reproduction, different sorts of cells are built. This question is important in speculating about the prebiotic emergence of self-organization in the soup of chemicals that used to be Earth. We're trying to figure out how self-organization occurred, how it bootstrapped Darwinian evolution, how DNA came out of that, and so on. The current dogma is that DNA is central, but DNA may have come along a lot later, as a regulatory mechanism.

In other computational experiments, we're looking at very simple animals and modeling their neural development. We're looking at polyclad flatworms, which have a primitive but adaptable brain with 2,000 or so neurons. If you take a polyclad flatworm and cut out its brain, it doesn't perform all its usual behaviors, but it can still survive. If you then get a brain from another polyclad flatworm and insert it into the brainless flatworm, after a few days it can carry out all of its behaviors pretty well. If you turn the brain 180 degrees and put it in backward, the flatworm will walk backward a little bit for the first few days, but after a few days it will be back to normal, with this brain helping it out. The brain adapts and regrows. How is that regrowth and self-organization happening in this fairly simple system? All these projects are looking at how self-organization happens.

Biological systems grow from simple to more and more complex. How do the mechanics of that growth happen? How does rigidity come out of fairly sloppy materials? On the computational side, I'm trying to build an interesting chemistry that is related to physics and has a structure in which you get interesting combinatorics out of simple components in a physical simulation, so that properties of living systems can arise through spontaneous self-organization. The question here is, What sorts of environmental influences do you need? In the prebiotic soup on Earth, you had tides, which were very important for sorting. You had regular thunderstorms every three or four days, which also served as regular sorting operations. You had the day-and-night cycle, heating and cooling. With this thermodynamic washing-through of chemicals, it may be that some clays attached themselves to start self-organization, but you had to get from crystal structure to this other sort of organization. What's the simplest chemistry you can have in which that self-organization will arise? What are its key properties? Obviously our chemistry had them.

We've come a long way since the early AI stuff. In the 1950s, when John McCarthy held that famous six-week meeting up in Dartmouth where he coined the term "artificial intelligence," people thought that the key to understanding intelligence was being able to solve problems that MIT and Carnegie Tech graduates found difficult. Allen Newell and Herbert Simon, for example, built programs that could start to prove some of the theorems in Russell and Whitehead's *Principia Mathematica*. Other people, like Alan Turing and Norbert Wiener, were interested in chess, something that people with a technical degree still find difficult. The concentration was all on intellectual pursuits. What these people missed was how important our own embodiment, our own perception of the world, is as the basis for our thinking. To a large extent, they ignored vision, which does much of the processing that goes on in your head. In our vision algorithms in robotics today, we can create things like face recognition and face tracking. We can do motion tracking very well now, actually. But we still cannot do basic object recognition. We can't have a system look at a table and identify a cassette recorder or a pair of eyeglasses, something a three-year-old can do. In the early days, such tasks were viewed as being too easy; everyone could do them, so no one thought they could be key. I used to carry around a 1966 memo—MIT Artificial Intelligence Memo #100—by Seymour Papert. He assigned an undergraduate a summer project of solving vision. He thought it would be easy, that an undergraduate would be able to knock it off in three months. It didn't turn out that way. Over time, there has been a realization that vision, sound processing, and early language may well be the keys to how our brain is organized, and that everything built on top of that is what makes us human and gives us our intellect. There's a whole new approach to creating intellectual robots, if you like, based on perception and language, an approach that was not there in the early days.

We're trying to generate some mathematical principles out of these robots and computational experiments. Principles, of course, are what we're really after, but my research methodology is not to go after something like that directly, because you can sit and twiddle your thumbs and speculate for years and years. I try to build real systems and then try to generalize from them. If we or others are successful in these attempts and can get to a real understanding of how the different pathways inside a living system interact to create that system, then a new level of technology can be built on top of that. We will then be able in a principled way to manipulate biological material just as we've learned in the last couple of centuries to manipulate steel and then silicon. Fifty years from now, our technological infrastructure and our bodies may be indistinguishable, in that they'll be the same sorts of processes.

MAKING MINDS

Hans Moravec

Perhaps programs that implement humanlike intelligence in a highly abstract way are possible on existing computers, as AI traditionalists imagine. Perhaps, as they also imagine, devising such programs requires lifetimes of work by world-class geniuses. But it may not be so easy.

HANS MORAVEC is a principal research scientist at the Robotics Institute of Carnegie Mellon University and the author of *Mind Children* and *Robot.*

Computers were invented in recent times to mechanize tedious manual informational procedures. Such procedures were themselves invented during the last ten millennia, as agricultural civilizations outgrew village-scale social instincts. Those instincts arose in our hominid ancestors during several million years of life in the wild, and were built on perceptual and motor mechanisms that had evolved in a vertebrate lineage spanning hundreds of millions of years.

Bookkeeping and its elaborations exploit ancestral faculties for manipulating objects and following instructions. We recognize written symbols in the way our ancestors identified berries and mushrooms; we operate pencils as they wielded spears; we learn to multiply and integrate by parts as they acquired village procedures for cooking and tentmaking. Paperwork uses evolved skills but in an unnaturally narrow and unforgiving way. Where our ancestors worked in complex visual, tactile, and social settings, alert to subtle opportunities or threats, a clerk manipulates a handful of simple symbols on a featureless field. And while a dropped berry is of little consequence to a gatherer, a missed digit can invalidate a whole calculation.

The peripheral alertness by which our ancestors survived is a distraction to a clerk. Attention to the texture of the paper, the smell of the ink, the shape of the symbols, the feel of the chair, the noise down the hall, digestive rumblings, family worries, and so on can derail a procedure. Clerking is hard work more because of the preponderance of human mentation it must suppress than the tiny bit it uses effectively.

Like little ripples on the surface of a deep turbulent pool, calculation and other kinds of procedural thought are possible only when the underlying turbulence is quelled. Humans achieve this quiescence, imperfectly, by intense concentration. Much easier to discard the abyss altogether: Ripples do better in a shallow pool. And numbers are better manipulated as calculus stones or abacus beads than in human memory. A few cogwheels in Blaise Pascal's seventeenth-century calculator performed addition better and faster than a human mind. Charles Babbage's nineteenth-century analytical engine would have outcalculated dozens of human computers and eliminated their errors. Such devices were effective because they encoded the bits of surface information used in calculation and not the millions of distracting processes churning the depths of the human brain.

But the deep processes sometimes help. We guess quotient digits in long division with a sense of proportion that our ancestors perhaps used to divide food among mouths. Mechanical calculators, unable to guess, plod through repeated subtractions. More significantly, geometric proofs are guided (and motivated!) by our deep ability to see points, lines, shapes, and their symmetries, similarities, and congruences. And true creative work is shaped more by upwellings from the deep than by overt procedure.

Calculators gave way to Alan Turing's universal computers and grew to thousands, then millions, and now approaching billions of storage locations and procedural steps per second. In doing so, they transcended their paperwork origins and acquired their own murky depths. For instance, minor operating system bugs may let one computer process spoil another, like a clerk derailed by stray thoughts. On the plus side, superhumanly huge searches, table lookups, and the like can sometimes function like human deep processes. In 1956 Allen Newell, Herbert Simon, and John Shaw's Logic Theorist's massive searches found proofs as a novice human logician would. Herbert Gelernter's 1963 Geometry Theorem Prover used large searches and Cartesian coordinate arithmetic to equal a reasonably talented human geometer's visual intuitions. Deep Blue's gigascale search, opening, and endgame books and carefully tuned board evaluations defeated the top human chess player in 1997.

Despite such isolated soundings, computers remain shallow pools. No reasoning program even approaches the sensory and mental depths habitually manifested at the surface of human thought. Many critics contrast computers' superiority in rote work with their deficits of comprehension and conclude that computers are prodigiously powerful but that computation lacks some human mental principle (of a physical, situational, or supernatural kind, according to your taste). Some artificial intelligence practitioners hold a related view: Computer hardware is sufficient, but difficult unsolved conceptual problems keep us from programming true intelligence. The latter premise seems plausible as far as reasoning is concerned, but it is preposterous for sensing. The sounds and images processed by human ears and eyes represent megabytes per second of raw data, enough to overwhelm even current computers. Text, speech, and vision programs derive meaning from snippets of such data by weighing and reweighing thousands or millions of hypotheses. At least some of the human brain works similarly. Roughly ten times per second at each of the retina's million effective pixels, dozens of neurons weigh the hypothesis that a static or moving boundary is visible then and there. The visual cortex's 10 billion neurons elaborate those results, each moment appraising possible orientations and colors at all the image locations. Efficient computer-vision programs require over 100 calculations each to make similar assessments. Most of the brain remains mysterious, but all its neurons seem to work about as diligently as those in the visual system. Elsewhere I've detailed the retinal calculation and concluded that it would take on the order of 100 trillion calculations per second of computing—about a million present-day PCs—to match the brain's functionality.

That number presumes an emulation of the brain at the scale of image-edge detectors, a few 100,000 calculations per second doing the job of a few 100 neurons. The computational requirements would increase (maybe a lot) if we demanded emulation at a finer grain—say, explicit representation of each neuron. By insisting on a fine grain, we constrain the solution space and outlaw global optimizations; however, by constraining the space we simplify the search! No need to find efficient algorithms for edge detection and other 100-neuron-scale nervous system functions. If we had good models for neurons and a wiring diagram of a brain, we could emulate it as a straightforward network simulation. The problems of artificial intelligence would be reduced to merely instrumentally and computationally daunting work.

Alternatively, we could try to implement the brain's function at much larger grain. The solution space expands, and with it the difficulty of finding globally efficient algorithms, but their computational requirements decrease. Perhaps programs that implement humanlike intelligence in a highly abstract way are possible on existing computers, as AI traditionalists imagine. Perhaps, as they also imagine, devising such programs requires lifetimes of work by world-class geniuses. But it may not be so easy. The most efficient programs exhibiting human intelligence might exceed the power and memory of present PCs manyfold, and devising them might be superhumanly difficult. We don't know. The pool is extremely murky below the ripples and has not been fathomed.

Each approach to matching human performance is interesting intellectually and has immediate pragmatic benefits. Reasoning programs outperform humans at important tasks, and many already earn their keep. Neural modeling is of great biological interest and may have medical uses. Efficient perception programs are interesting to biologists and useful in automating factory processes and data entry. But which method will succeed first? The answer is surely a combination of all those techniques and others, but I believe the perception route, currently an underdog, will play the largest role.

Reasoning programs are superb for consciously explicable tasks but become unwieldy when applied to deeper processes. In part, this is simply because the tasks deep in the subconscious murk elude observation, but also the deeper processes are quantitatively different. A few bits of problem data ripple across the conscious surface, but billions of noisy neural signals seethe below. Reasoning programs will become more powerful and useful in coming decades, but I think that comprehensive verbal common sense, let alone sensory understanding, will continue to elude them.

Entire animal nervous systems—hormonal signals and interconnection plasticity included—may become simulable in coming decades, as imaging instrumentation and computational resources rapidly improve. Such simulations will greatly accelerate neurobiological understanding, but I think not rapidly enough to win the race. Valentino Braitenberg, former director of the Max Planck Institute for Biological Cybernetics, who analyzes small nervous systems and has designed artificial ones, notes the rule of "downhill synthesis and uphill analysis": It is usually easier to compose a circuit with certain behaviors than to describe how an existing circuit manages to achieve them. Meager understanding and thus means to modify designs, the cost of simulating at a very fine grain, and ethical hurdles as simulations approach the human scale—all these will slow the applications of neural simulations. No human-scale intelligence, as far as we know, ever developed from conscious reasoning down or from simulations of neural processes, and we really don't know how hard doing either may be. But the third approach is familiar ground.

Multicellular animals with cells specialized for signaling emerged in the Cambrian explosion a half-billion years ago. In a game of evolutionary one-upmanship, maximum nervous-system masses doubled about every 15 million years, from fractional micrograms, then to several kilograms now—albeit with several abrupt retreats (often followed by accelerated redevelopment) when catastrophic events eliminated the largest animals. Our gadgets, too, are growing exponentially more complex, but 10 million times as fast as that. Human foresight and human culture move things along faster than blind Darwinian evolution. The power of personal computers has doubled annually since the mid-1990s; today's PCs might be comparable only to the milligram nervous systems of insects or the smallest vertebrates (e.g., the one-centimeter dwarf goby fish), but humanlike power is just thirty years away. A sufficiently vigorous development with well-chosen selection criteria should be able to incrementally mold that growing power in stages analogous to those of vertebrate mental evolution. I believe that a certain kind of robot industry will do this naturally. No great intellectual leaps should be required. When insight fails, Darwinian trial and error will suffice. Each ancestor along the lineage from tiny first vertebrates to ourselves became such by being a survivor in its time, and similarly ongoing commercial viability will select intermediate robot minds.

Building intelligent machines by this route is like slowly flooding puddles to make pools. Existing robot control and perception programs seem like muddy puddles, because they compete in areas of deepest human and animal expertise. Reasoning programs, though equally shallow, comparatively shine by efficiently performing tasks that humans do awkwardly and animals not at all. But if we keep pouring, the puddles will surely become deeper. That may not be true for reasoning programs: Can pools be filled surface down?

Many of our sensory, spatial, and intellectual abilities evolved to deal with a mobile lifestyle; an animal on the move confronts a relentless stream of novel opportunities and dangers. Other skills arose to meet the challenges of cooperation and competition in social groups. Elsewhere I've outlined a plan for commercial robot development that provides similar challenges. It will require a large, vigorous industry to search for analogous solutions. Today the industry is tiny. Advanced robots have insectlike mentalities, besting human labor only rarely, in exceptionally repetitive or dangerous work. But I expect a mass market to emerge in this decade. The first widely usable products will be guidance systems for industrial transport, and cleaning machines that three-dimensionally map and competently navigate unfamiliar spaces and can be quickly taught new routes by ordinary workers. I have been developing programs that do this. They need about a billion calculations per second, like the brainpower of a guppy. Industrial machines will be followed by mass-marketed utility robots for homes. The first may be a small, autonomous robot vacuum cleaner that maps a residence, plans its own routes and schedules, keeps itself charged, and empties its dustbag when necessary into a container. Larger machines with manipulator arms and the ability to perform several different tasks may follow, culminating eventually in human-scale "universal" robots that can run application programs for most simple chores—programs that their 10-billion-calculation-per-second, lizard-scale minds will execute with reptilian inflexibility.

This path to machine intelligence—incremental, reactive, opportunistic, and market-driven—does not require a long-range map, but it has one in our own evolution. In the decades following the first universal robots, I expect a second generation with mammal-like brainpower and cognitive ability. These robots will have a conditioned learning mechanism and steer among alternative paths in their application programs on the basis of past experience, gradually adapting to their special circumstances. A third generation will think like small primates and maintain physical, cultural, and psychological models of their world to mentally rehearse and optimize tasks before physically performing them. A fourth, humanlike, generation will abstract and reason from the world model. I expect that the reasoning systems will be adopted from the traditional AI approach maligned earlier in this essay. The puddles will have reached the ripples.

Robotics should become the largest industry on the planet early in this evolution, eclipsing the information industry. The latter achieved its exalted status by automating marginal tasks we used to call paperwork. Robotics will automate everything else!

QUANTUM COMPUTATION

DAVID DEUTSCH

For me, the main application of the theory [of quantum computation] *is to change our sense of the nature of reality. Regardless of its practical applications in the distant future, the really important thing is the philosophical implications— epistemological and metaphysical—and the implications for theoretical physics itself. One of the most important implications is one that we get before we even build the first qubit* [quantum bit]. *The very structure of the theory forces upon us a view of physical reality as a* multiverse.

DAVID DEUTSCH's papers on quantum computation laid the foundations for that field, breaking new ground in both physics and the theory of computation and triggering an explosion of research efforts worldwide. His work revealed the importance of quantum effects in the physics of time travel, and he is the most prominent contemporary researcher in the quantum theory of parallel universes. In 1998, he was awarded the Paul Dirac Prize by Britain's Institute of Physics "for pioneering work in quantum computation leading to the concept of a quantum computer and for contributing to the understanding of how such devices might be constructed from quantum logic gates in quantum networks." He is a founding member of the Centre for Quantum Computation at the Clarendon Laboratory, University of Oxford, and the author of *The Fabric of Reality*.

───────────

My questions are in the direction of even deeper connections between physics and the theory of computation. We have to regard the Turing theory—the traditional theory of computation—as being just a classical approximation to the real, quantum theory of computation. We already know of a few issues in theoretical physics which can usefully be regarded as computational questions—questions about how information can or cannot be processed. One thing I am aiming for is a new *kind* of theory—quantum constructor theory, which is the theory of what can be built or, more generally, the theory of what can be done physically.

What sorts of computations do physical processes correspond to? Which of those computations can be realized with what resources, and which sorts can't be realized at all? What little we know about this new subject consists of a few broad limitations, such as the finiteness of the speed of light. The theory of computability and complexity theory give us more detail on the quantum side. But the big, unanswered technological question in my field at the moment is, Can *useful* quantum computers actually be built? The basic laws of physics permit them, as far as we know. We can design them in theory; we know what physical operations they would have to perform. But there is still room for doubt about whether we can build them out of actual atoms and make them work in a useful way. The debate is not even a scientific one at the moment, because there is no scientific theory about what can and can't be built.

Similar questions are raised by the whole range of nanotechnology that has been proposed in principle. So that's where a quantum constructor theory is needed. It's needed because quantum theory is our basic theory of the physical world. All construction is quantum construction. Quantum computing is information processing that depends for its action on some inherently quantum property, especially superposition. Typically we would superpose a vast number of different computations—potentially more than there are atoms in the universe—and bring them together by quantum interference to get a result. Apart from quantum cryptography, this research is unlikely to have many practical applications in the near or medium-term future; nevertheless, it does give us some immediate benefits. Here's a recent example from my own work.

Quantum mechanics, in the traditional formulation, seems to have a nonlocal character—that is, things you do *here* seem instantaneously to affect things that happen *there*. It has been known from the beginning that this so-called nonlocality cannot be used to send signals—that is, information. But still, philosophically, what are we to make of it? What sort of reality is quantum mechanics telling us we live in? And of course it's hard not to wonder, "Well, if *something* gets there instantaneously, that something is going faster than light. So in another reference frame, it's traveling into the past. So it could create paradoxes; couldn't that solve the problem of consciousness, explain telepathy, summon up ghosts?"—you name it. This "nonlocality" is one of the ideas that has fueled the appalling mysticism and doubletalk that has grown up around quantum mechanics over the decades.

But once you understand that the idea is all about information processing, it becomes much easier to stop the handwaving and start calculating where information actually goes in quantum phenomena. That's what my colleague Patrick Hayden and I did ["Information Flow in Entangled Quantum Systems," *Proc. R. Soc. Lond.*, A456, 1759-1774, 2000]. The results blow the quantum nonlocality misconception clean out of the water. Doing things *here* can affect things *there*—visibly or invisibly—only when the information about what you've done here has traveled there in some information-carrying physical object. There is nothing instantaneous, nothing nonlocal, nothing mystical.

Experiments that supposedly demonstrate quantum nonlocality in the lab don't really do that. They demonstrate quantum *entanglement*—one of the fundamental quantum phenomena, but a local one. It turns out that when it looks as though there's a nonlocal effect, what's really happening is that some of the information in quantum objects has become inaccessible to direct observation. In our analysis, we actually track how this information travels during entanglement phenomena. It never exceeds the speed of light, and it always interacts in a purely local way. By the way, the presence of such not-directly-accessible information can be seen as the very thing responsible for the power of quantum computers. The insights we gained from that work are leading in other very promising directions, too.

I am currently working on two spin-offs of that paper. One is work on the structure of the "multiverse"—making precise what we mean by such previously hand-waving terms as "parallel universes." It turns out that the structure of the multiverse is largely determined by the flow of quantum information within it, and I am applying the techniques we used in that paper to analyze that information flow. The other is a generalization of the quantum theory of computation to allow it to describe exotic types of information flow such as we expect to exist in black holes and at the quantum gravity level.

This is all in the context of my ever-strengthening conviction that the quantum theory of computation *is* quantum theory. The quantum theory of computation provides the clearest and simplest language and mathematical formalism for setting out quantum theory itself. I think that quantum mechanics textbooks will soon begin using quantum computations as their introductory examples, rather than calculating the energy levels of the hydrogen atom and suchlike—examples that contain a high proportion of irrelevant stuff. Quantum computation gets down to basics because quantum computation *is* the basics.

For me, the main application of the theory is to change our sense of the nature of reality. Regardless of its practical applications in the distant future, the really important thing is the philosophical implications—epistemological and metaphysical—and the implications for theoretical physics itself. One of the most important implications is one that we get before we even build the first qubit [quantum bit]. The very structure of the theory forces upon us a view of physical reality as a multiverse. Whether you call this "the multiverse" or "parallel universes" or "parallel histories" or "many histories" or "many minds"—there are now half a dozen or more variants of this idea—what the theory of quantum computation does is force us to revise our explanatory theories of the world, to recognize that the world is much bigger than it looks.

When I use the term "bigger," I am getting at the following: Suppose we were to measure the sizes of things in terms of the amount of information needed to describe them. To specify the positions of the atoms in a room, I need three numbers for each atom. The more atoms I want to describe, the more numbers I need. The more accurately I want to do it, the more decimal places I need to give. I can think of doing that for the whole universe. That may sound a lot of information, because there are 10^{80} atoms in the known universe, not to mention the other degrees of freedom. So the amount of information may seem unimaginably vast. Yet it is minuscule compared with the amount of information needed to specify the computational state of a single quantum computer sitting on some future laboratory bench. So in conceptual terms a quantum computer is a much bigger object than the whole of the classical universe. This fact forces quite a change in our worldview.

A quantum computer would be an object far more complex than the whole of the classical universe. The whole of physical reality is like that, too, of course, and we sometimes call it the multiverse. We see a classical universe out there because most of the multiverse is not directly accessible. You can infer the existence of hidden quantum information only indirectly, as in the entanglement experiments I mentioned.

To many people, this conclusion was compelling even before quantum computers. The many-universes interpretation was proposed in 1957. But you can construe all the earlier arguments as being computational arguments, too. The people making them didn't think of them as such, but that's what they were. They were saying, "We look around us and we see something that's approximately a classical universe, and we expect that if you take quantum mechanics into account it might add a certain amount of extra 'stuff'—just as relativity did—which behaves differently, but there's still roughly the same amount of reality as we thought there was." But that's *not* what happens when you take quantum mechanics into account. When you take quantum mechanics into account, reality becomes vastly, exponentially bigger and more complex than it was under classical physics.

If the system is a quantum computer, we can tell there is "hidden information" in it because of the answers it gives us. Take Grover's quantum search algorithm, for instance. It works like this: Let's say you're trying to crack a code—to guess the secret key, say. You're searching through all the possible keys. It is a trivial theorem of classical computation that if you want to search through a trillion unknown things, you generally have to do a trillion physical operations of some kind. You might be able to do some of them in parallel, but a given computer will only be able to do a fixed number at a time in parallel. One way or another, you have to do a trillion things, so if you want to use the same computer to search through 2 trillion things it must take at least twice as long, and so on.

But with a quantum computer, you can do better. First of all, to search through a list of a trillion things, you need do only a million operations. In general, in order to search through n possibilities, one need do only the square root of n physical operations. And then, if you let your quantum code breaker think for twice as long, it will examine *four* times as many keys. Three times as long, *nine* times as many, and so on. The explanation of this, in terms of many universes, is very simple. It's just that there are the square root of n universes collaborating on such a task. But again, never mind the question of interpretation as such. If we just think of what this computation implies for the reality we find ourselves in, again, the answer is that reality is much bigger than it looks. Finding the right key depended, logically, on searching all those other keys. So, as a matter of logic, those other possible keys must all have existed somewhere and been checked by something for whether they fitted.

Ultimately, information has to have a physical realization; that's why it does come down to atoms or stars or whatever in the end. But because of the universality of computation, you don't have to think in terms of specific implementations. I don't have to know whether my information is going to be stored on a magnetic disk or whatever; I just know that more information means a bigger object.

If computers are going to continue to become more powerful, processors and memory devices must become smaller. For that reason alone, quantum processes must be harnessed. Whether to make quantum computers or not doesn't really matter. Even to make classical computers out of atomic-scale components, you'd have to use quantum physics and, ultimately, the quantum theory of computation. And once you're making those, the same technology could probably also make quantum computers. And the incentive would be there, because of the various inherent advantages of quantum computation.

Proposed technologies for building them are at present competing. We don't know which way the competition is going to go. It could be ion traps or it could be quantum dots, or other solid-state devices, or it could be superconducting loops. It could be molecules, or something we don't know about yet.

At present, the biggest quantum computer in the world has about three qubits. Not much practical use, and it requires quite a large apparatus to make it work. Yet with three qubits, you can already implement quantum algorithms that no classical computer using three bits could mimic.

Quantum cryptographic devices already exist in the laboratory. Eventually that's going to give perfectly secure communication. No longer will cryptography depend on the difficulty, or the intractability, of guessing an unknown key. It will simply be physically impossible to discover the key if you don't have the relevant physical object; that is the ultimate in cryptography. The trouble is that right now quantum cryptography is severely limited in range. It can't be done through open air. It's got to be done through fiber-optic cable, and I think the world record is about 100 kilometers. But still, you could wire up the city of London, or central Washington D.C., with absolutely secure communications. I don't know why that hasn't been done. I doubt that it has anything to do with sinister machinations by the government, though; it's probably just that it takes a long time for an idea to become genuinely commercially viable. We already know how to build absolutely secure communications if we want to, at ranges of a few kilometers. Longer ranges would present a problem, but at least one group at Los Alamos is working on a system that will allow us to bounce quantum-encrypted messages off a satellite, and that would essentially solve the problem. In the long run, the problem could also be solved by quantum repeating stations. Unfortunately they would require much more sophisticated quantum computation than the raw cryptography does. They will come along eventually, perhaps in a decade or two.

Another thing that will come along—probably after more than a decade or two—is using a quantum computer to decrypt existing codes, as I described just now. Quantum decryption machines would render existing cryptographic systems obsolete. I've been surprised repeatedly by how well the experimentalists have been able to implement theoretical concepts in quantum computing. Apart from quantum cryptography, I'd be amazed if quantum computing produces anything technologically useful in ten years, twenty years, even longer. But I've been amazed before.

WHAT COMES AFTER MINDS?

MARVIN MINSKY

Tens of thousands of researchers today, in the field called artificial intelligence, are striving to endow machines with . . . humanlike abilities. They've developed programs that outperform people in many specialized domains. Some solve hard mathematical problems or skillfully pilot ships and planes. Others can recognize voices and faces or objects on assembly lines. But none of them yet can dress themselves, or understand the sorts of things that young children can. Why don't any computers yet have what we call everyday, common-sense knowledge or do the sorts of reasoning that we regard as obvious?

MARVIN MINSKY is Toshiba Professor of Media Arts and Sciences and professor of electrical engineering and computer science at MIT. His research has led to both theoretical and practical advances in mathematics, computer science, physics, psychology, and artificial intelligence, with notable contributions in the domains of computational semantics and knowledge representation, machine perception and learning, and theories of human problem solving. Minsky is also the inventor of the popular Confocal Scanning Microscope, which revolutionized our ability to see dense microscopic structures. He is the author of *The Society of Mind* and *The Emotion Machine*.

———————

Why has it been hard to find out how minds work? The more we can learn about how human minds work, the better we will be able to guide the development of our genetic successors or of beings whose making we'll supervise. But why should we alter ourselves instead of forever remaining the same? That's because we have no alternative. If we stay unchanged in our present state, we are unlikely to last very long—on either cosmic or human scales of time. In the next hundred or thousand years, we are liable to destroy ourselves, yet we alone are responsible not only for our species' survival but for the continuation of intelligence on this planet, and quite possibly in this universe. For us to develop our future potential, we'll have to protect our environment from both climatic warming and glaciation. We also must keep from being extinguished by other kinds of accidents, like the collisions with comets or asteroids that more than once destroyed most species in the past. We're also aware that our sun will engulf us in no more than another few billion years—and possibly quite a bit sooner than that. We've seen many suggestions about dealing with these, but none of them yet seem practical. A more practical course might simply be to think less about those issues themselves and focus instead on finding ways to make ourselves more intelligent!

Psychology still is in its infancy. We find many ideas in Aristotle's *Rhetoric*, written twenty-three centuries ago, about how thinking works, and much of that text still seems up-to-date, which is not true of most other sciences. This suggests that we ought to closely examine why far more advanced ideas about minds did not begin to appear until late in the nineteenth century—for example, until the research of thinkers like Wilhelm Wundt, Francis Galton, William James, and Sigmund Freud. And even then, progress was slow. What kept psychology from advancing more quickly? Here are some possible reasons:

- *The "Single Self" model.* A principal obstacle to progress was the "Single Self" model of the mind—that is, the common-sense notion that each of us has a single, central identity that has definite intentions and goals. The trouble is that this idea by itself tends to keep you from thinking about what minds are and how they work. Freud was one of the first to challenge this model, by suggesting an architectural theory, in which the mind consists of a number of systems—with "thinking" resulting from the ways they conflict.

- *Lack of good ways to describe information.* Mental processes clearly deal with symbols, whose "meanings" refer, at least some of the time, to self-reflections about those same processes. In older times, we had no ways to work with or represent systems like these—and no such techniques were to arise till the dawn of computer science.

- *Complex systems and physics envy.* Most early psychologists were so impressed by the progress of physics that they kept trying to emulate Maxwell or Newton by looking for very small sets of laws to account for human behavior. This constraint has often been reified under the name of Occam's Razor: Never assume any entities that do not seem logically necessary. However, we learned, even from early neurology, that our brains have hundreds of different mechanisms. This suggests looking for theories with more parts, not fewer.

- *Common-sense thinking.* Modern psychology has made great progress toward understanding human perception and simple reactions. However, we still have not tried hard enough to unravel the mysteries of how we represent common-sense knowledge, or how we use the knowledge we have for solving difficult problems.

THE SINGLE SELF VS. THE FREUDIAN MODEL

Imagine a child who wants a toy that some smaller youngster is playing with. The most obvious solution is to take it by force, but our cultural values prohibit this and our child has to deal with that prohibition. However, a conflict is hard to represent in the model of a Single Self, because it is difficult to envision how a single mind could entertain several conflicting ideas.

Freud tackled this problem by viewing the mind as having several almost-separate parts, each with its own machinery. In a simplified version of his idea—which I think of as the "Freudian Sandwich"—one begins (like other animals) with appetites, impulses, urges, and drives, embodied in an inborn system that Freud called the Id. But we also grow up in a social world from which we acquire additional goals, of the kinds that we sometimes call ideals. Freud envisioned these as embodied in a second system, the Superego. Then he went on to describe the rest of the mind as a highly diverse assortment of schemes: the collection he called the Ego. He saw that system (which includes what we call common-sense reasoning) as sandwiched between those other two systems; its job is to find acceptable ways to settle the conflicts that so often arise between our instinctive goals and our highest ideals.

We could also interpret Freud's idea as suggesting that much of what the mind does is involved with what we now call "debugging." Because our behavior is not based on simple laws like those of a neat mathematical theory, our minds work more like large bundles of software, each part of which suffers from various bugs. (This is not only the case with brains; it is generally what evolution yields. Instead of each subsystem being perfect, older systems get patched together by the addition of other systems, each of which helps to fix some earlier bugs and thereby creates additional ones.) In particular, Freud imagined that some parts of the mind are like monitors that watch other parts and learn how, when those go wrong, to suppress them and turn on others.

We don't have to accept Freud's particular theories; indeed, he kept changing them over the years. But we ought to recognize the significance of his approach: It offered alternatives to the view that the mind needs a central "person" or "soul." Instead, we can see the mind as a collection of

structures that can both cooperate with and oppose one another to find ways to deal with conflicting goals. In fact,
Freud's theory had more than three parts; he also envisioned
all sorts of mental critics, censors, and suppressors, as well as
multiple ways to make representations. However, there still
were no ways, in those early times, to make good descriptions of such things; that had to wait for more modern ideas
about the representation of knowledge and also better ways
to represent processes.

Thus, instead of searching for uniform rules that apply to
all our mental functions, this "multiple agent" view makes
room for many more kinds of resourcefulness, because each
of those partially separate systems can work in accord with
different laws. To be sure, that insight alone doesn't help, because we need to know how the systems are organized.
Freud's ideas should have led more psychologists to pursue
the higher-level problem of describing the architectures of
minds. Freud's followers have done some of that, but not a
lot has come from it, because those ideas did not much affect
the other main streams of psychology. I suspect that this may
have been because of what I described as physics envy; few
scientists could envision ways to bridge the gap between the
simple and popular rule-based theories and these lofty but
vague architectural schemes. I recently asked a large class of
students, most of whom had taken courses in "cognitive
science," about what they knew of Freud's ideas, and few
recalled any mention of them. In effect, Freud's ideas have
been ostracized, perhaps because in earlier years they
were seen as politically incorrect. Besides, there was no feasible way to predict how such a large system would behave;
that had to wait until big computers arrived.

As for modern computer programs, their various parts
sometimes get into conflicts, but our programmers tend to

regard this as not an acceptable way to proceed. Instead, they prefer to try to find ways to make each program work perfectly. I'm not saying that this is a bad idea—only that it will never completely succeed, so we'll have to develop alternatives. One reason that our mammalian brains have so many different specialized "centers" must be that as our ancestors evolved, their brains had to develop new mechanisms to adapt to new ecological niches, whereas most other animals failed to evolve multiple different "ways to think"—which is why a species typically can survive only in a single kind of environment, its ecological niche. In contrast, our brains became a great jumble of stuff, and psychology has yet to recognize this with theories about how thinking can work in spite of it or by finding ways to exploit it.

Most present-day computer programs still resemble those typically specialized animals: If you type even a slightly wrong command, such a program will quickly die, whereas if you say inconsistent things to a person, you are likely to get responses like "That's funny" or "I don't believe you" or "I'll try to understand your point of view." Each person has many different ways to deal with each difficult situation. My thesis in *The Emotion Machine* is that *no* uniform scheme will lead to machines as resourceful as a human brain. Instead, I'm convinced that this will require many different "ways to think"— along with bodies of knowledge about how and when to use each of them.

THE SIGNIFICANCE
OF COMPUTER SCIENCE

In the middle twentieth century, something happened that I expect will eventually transform our civilization: the development of computer science. Most people assume that "computer science" refers to the kinds of things that computers do, but this assumption ignores its importance. I see computer science as scarcely about computers at all, but instead as a radically new collection of ways to depict and (thereby) think about systems that are extremely complex.

By the way, I'm not using the term "complex" to mean any system that has a great many parts. We're concerned with special cases in which those parts interact in nonuniform ways. Thus a human brain has evolved some hundreds of parts, each of which has different behaviors.

Such inhomogeneous systems do not like to submit to the kinds of theories that worked so well in fields like physics and mathematics. Mathematics is a collection of methods that deal well with systems that are based on simple principles—no matter that their external behaviors may seem enormously complex at first. Thus mathematics can sometimes deal with systems that have enormous numbers of parts, but only when these parts interact in ways that can mostly be ignored! For example, statistical mechanics is good at explaining some properties of very large systems whose parts all have similar properties, but not when too many of those parts are different. At another extreme, "chaos theories" can sometimes help us explain why some apparently simple systems can produce complex behavior, when small differences can lead to exponentially growing changes. Other mathematical methods can sometimes explain (in principle) how some complex systems produce simple behavior. However, such theories rarely help us to understand the details of specialized, complex systems whose behaviors turn out to be *useful* to us—such as programs based on "conditionals" that must be described in terms of "ifs" and "thens." In those kinds of systems, small differences can *instantly* cause drastic behavioral changes—for example, of the kind that suddenly happens whenever we get a new idea!

In contrast, computer science provides us with huge collections of new useful concepts that can help us describe mental processes. For example, most early theories of memory suggested that knowledge is stored as simple connections between separate items—or, even more simply, like propositions stored in a box. Computer science has helped us envision a far wider range of ways to represent different types and forms of knowledge, such as:

- items in a database,
- connections within a neural net,
- sets of "if/then" reaction rules,
- structures linked into semantic networks,
- programlike procedural scripts,
- interconnected collections of frames,
- hash-coded recognition schemes,
- multiple levels of memory cache, etc.

Our minds depend on very large networks of different sorts of processes, and I expect us to discover that they use many specialized representations. Only new ideas like these about representing information provide us with adequately specialized ways to describe such things. To be sure, our programmers had to pay a high price for the convenience of using such intricate schemes; they left us bereft of the certainty that comes with neat mathematical proofs. In exchange, though, we gained the use of what AI researchers call heuristic knowledge—that is, knowledge about which processes will usually help to solve each particular type of problem. Also the use of computers enabled us to simulate what such systems do, which provides us with a (sometimes inadequate) substitute for mathematical proof.

CAUSES OF HUMAN RESOURCEFULNESS

A typical modern computer program can solve only one particular type of problem and only in one particular way. In contrast, a person who gets stuck when using one method can usually switch to some other approach. Furthermore, when you change your technique, you usually don't have to start the process all over again; instead you change your tactics and representations and keep on going from where you were.

How can you change how you think about things without needing to start all over again?

- *We often use multiple representations.* One answer to that question would suggest that whenever we learn from experience, we usually make several representations of each thing we've learned. To do this, we build several different memory schemes, each of which serves different purposes or helps with different problem types. For example, when you encounter a new kind of object, you can represent it as a network of visual, aural, and tactile descriptions—and you usually go even further than that, because various parts of your mind ask questions like "Who owns that object?" or "Why is this here?" or "How much does it cost?" or "How does it work?" I suspect that you usually also ask, "What could I possibly use it for?" or "How can I gain control over it?" Then each of these questions may lead you to store yet additional types of descriptions of the object, along with appropriate interconnections, so that when you later ask different questions you can quickly change your manner of thinking.

- *Emotions are different ways to think.* Our common-sense "folk psychology" tends to regard "thinking" as relatively simple, whereas we like to think of "emotions" as extremely complex and mysterious. Thus thinking, in that popular view, consists of scarcely anything more than cold, mechanical, logical processes that are not especially interesting. In contrast, emotions are seen as completely different, with all their unpredictable colors and inexplicable feelings. Consider an analogy between the colors and shapes of physical objects and the feelings and thoughts that come with ideas. We don't usually see any mystery about the physical shape of a thing, because we can describe it in terms of smaller parts, along with their spatial relationships. But we don't have the same view of colors, because they seem utterly different from shapes—quite separate and additional. Indeed, colors seem so detached and "accessory" that we have virtually nothing to say about them. (*Why* do colors seem so different from shapes? I suspect that this could be simply because they're related to processes inside our brains that evolved more recently, hence are less well connected to the rest of our thoughts.) However, the view proposed in *The Emotion Machine* takes an almost opposite view: Emotions are not additions to thoughts. Instead, I see each emotional state as a distinctly different way to think. Indeed, I'd argue that in many cases an emotional state does not result from adding to thought; instead, it may come from suppressing resources that one otherwise usually uses when thinking! For example, when something makes you angry enough, you may start to suppress some of your longer-range plans, shut off some mechanisms of defense, and limit the extent of your thinking to more shallow and less self-reflective levels. In other words, that emotional state results in a manner of thinking that comes in large part from shutting off some of your usually active resources.

- *Thinking involves vast collections of knowledge.* What makes people so much smarter than most other species of animals? Clearly this is partly because we learn more—both more knowledge about particular things and better ways to think about them. Also, we learn such things on multiple levels: We learn not only new ways to think but also about when and how to use them.

However, we have not yet learned very much about how to make our machines learn to do all this. Tens of thousands of researchers today, in the field called artificial intelligence, are striving to endow machines with such humanlike abilities. They've developed programs that outperform people in many specialized domains. Some solve hard mathematical problems or skillfully pilot ships and planes. Others can recognize voices and faces or objects on assembly lines. But none of them yet can dress themselves, or understand the sorts of things that young children can. Why don't any computers yet have what we call everyday, common-sense knowledge or do the sorts of reasoning that we regard as obvious?

I believe that it's largely because only a handful of researchers have tried to make theories about how machines could do common-sense thinking. What are those thousands of other AI researchers doing instead? My impression is that they have been afraid to make a frontal attack on this problem, and have tried instead to extend the traditional methods that have met with success on some specialized problem. The result has been a collection of fads that I see as futile attempts to solve problems; however, common-sense thinking is too complex for those older methods to work well enough.

Here are a few of those other attempts, each of which has made progress on some kinds of problems but has failed to develop more general methods:

- *Statistical models.* How do you understand a typical sentence, in which each word may have several meanings? One common approach is to use some statistics. Suppose that you read, "John picked up his pen." Now, that pen might be (a) a thing to write with or (b) an enclosure for a pig or (c) the internal horny shell of a squid. If we know nothing else about John, then (a) would be more likely than (b), which would be more likely than (c). If you knew that John was a biologist who frequently works with cephalopods, then (c) might have a fighting chance. If you knew that John was a farmer, then you might pick (b) because that has more correlation, but you'd also have a conflict, because most pigpens are too big for one person to lift. The use of language statistics leads to more or less what one ought to expect. As we enlarge the body of evidence, it will lead to increasingly more correct choices, but will eventually approach some limits at which alternative meanings cannot be distinguished, because they depend on much larger-scale contexts—for example, on references to other texts.

- *Insectlike robots.* Perhaps the most wasteful fad of all is the building of simple, small real-world robots—possibly based on the idea that an animal must "learn to crawl before it can walk" (a popular saying that's simply not true!). Thus today, at a host of universities, you'll find people building robots that can learn to walk through a field of obstacles or win simple games against competitors.

However, as far as I can see, thousands of such experiments have failed to produce new, important ideas that could not have been arrived at by more thoughtful reflection.

We've seen quite a few other approaches in recent years, such as neural networks, rule-based expert systems, Bayesian learning, Markoff models, predicate logics, complexity theories, and so on. Most of these schemes are more or less based on the concept that all one needs is a large enough computer. I won't review those systems here, except for genetic programming, which could become the most promising one, if we can overcome its deficiencies.

- *Genetic programming.* Here the basic idea is simply to simulate Darwinian evolution: Begin with some particular program, and if it does not solve your problem, make one or more mutations in it and try the experiment again. More generally, it may also help to make a large population of such programs and arrange competitions between them. This idea is tempting for quite a few reasons, most notably because this is how people evolved, so we know that it can in principle produce some wonderful things. In any case, the idea is attractive because it suggests a way to solve problems without any effort at planning and thinking. Many practitioners like to describe it as a fresh and novel approach, but to me it looks like a born-again version of what the earliest AI researchers tried and abandoned because it was much too slow. Today, using computers that are millions of times faster, these search schemes work quickly on some kinds of problems. But so (most practitioners haven't noticed) would some of those old-fashioned methods. The hope, of course, is that by making such programs larger, they could grow to solve much harder problems—and this idea, too, has spread over the world, engaging thousands of students and other people. But here I'll briefly try to point out that although Darwinian evolution is "natural," it also has some serious flaws.

- *Our genomes store no explicit goals.* First, Darwinian evolution has no place in which to place goals—and, accordingly, no place for subgoals, either. This means it is not well equipped to split hard problems into parts and then use "divide and conquer" methods. Because of this deficiency, evolution cannot exploit those techniques that otherwise could reduce the size of an exponentially growing search. (And, in fact, our biological systems have taken some steps toward doing this, through the invention of such mechanisms as homeoboxes and other elaborate controls over gene expression.) If animals had first evolved explicit ways to represent goals, then some steps that had to wait millions of years might have occurred in just a few generations. It took several hundred millions of years for us to evolve from our ancestor yeasts; perhaps a more stratified evolutionary scheme could have reduced this period by orders of magnitude.

    We may be on the verge of inventing some such schemes ourselves. Few programs, too, represent their goals, except perhaps as comments in their source codes; however, a class of programs was developed around 1960 by Allen Newell, Clifford Shaw, and Herbert Simon which did have explicit goals and subgoals and solved some significant problems. They called it the General Problem Solver; regrettably, few programmers today have heard of it, and are constantly reinventing it but in far less clear and powerful forms.

- *Our genomes don't represent records of failures.* Second, Darwinian evolution selects the animals that survive but has no explicit ways to remember what caused the nonsurvivors to die! So it is only able to learn how to deal with the most common kinds of mistakes. Thus, we could expect a species like mice to develop specific behaviors that help to protect them from, say, cats and snakes. However, no species can deal, genetically, with enormous numbers of uncommon mistakes. Of course, you could argue that our genes "remember," but genes only store tricks that got things to survive—with no records at all of the reasons the tricks succeeded. In particular, our genes have no way to remember large numbers of rare mistakes, so animals cannot inherit knowledge about large numbers of errors. This is because an animal has only thousands of genes. But a large enough brain can learn many millions of memes, their cultural equivalents. The growth of "intelligence" in a human individual depends in large part on learning a large body of common mistakes to avoid—if it grows up immersed in a culture that has effective ways to pass on such things. This way, each generation can pass on great catalogs of the awful mistakes that caused other people to die. Whenever you make a significant error, you can remember to never repeat it—and you can tell your friends—and thus our cultures can grow, because new generations don't have to start all over again.

COMPUTERS AND COMMON SENSE

A computer today is a million times more powerful than those of thirty years ago, but the programs and systems that they run have not changed that much, at least in certain important respects. Indeed, we're now seeing a large-scale reversion from systems like Windows—because they're too rigid and hard to maintain—back to a simpler system called Unix, designed in 1969. However, this is not the main reason why computers do not seem to have changed.

The main thing that has remained the same is that computers still know so little about their world. In particular, they have no ideas about the goals of the people who use them. This is why, for example, most programs will die whenever their users make a mistake, be it a grave conceptual error or just typing an incorrect character. Someday, however, computers will have the sorts of common-sense knowledge that most of us share—millions of everyday facts about the world and common-sense ways to think about them. There are some projects aimed toward that, but instead of reviewing the primitive state of our programs, let's envision what might happen if they succeed. If they learn to think about themselves and invent new ways to improve themselves, then everything we know will change and (if we can keep control of them) we'll never need to work again.

THE SINGULARITY

RAY KURZWEIL

We are entering a new era. I call it the Singularity. It's a merger between human intelligence and machine intelligence which is going to create something bigger than itself. It's the cutting edge of evolution on our planet. One can make a strong case that it's actually the cutting edge of the evolution of intelligence in general, because there's no indication that it has occurred anywhere else. To me that is what human civilization is all about. It is part of our destiny, and part of the destiny of evolution, to continue to progress ever faster and to grow the power of intelligence exponentially.

RAY KURZWEIL, an inventor and entrepreneur, has been pushing the technological envelope for years in his field of pattern recognition. He was the principal developer of the first omni-font optical character recognition machine, the first print-to-speech reading machine for the blind, the first CCD flat-bed scanner, the first text-to-speech synthesizer, the first music synthesizer capable of recreating the grand piano and other orchestral instruments, and the first commercially marketed large-vocabulary speech recognition system. In 1999 he received the National Medal of Technology from President Clinton. In 2002 he was inducted into the U.S. Patent Office's National Inventors Hall of Fame. He is the author of *The Age of Intelligent Machines*, *The Age of Spiritual Machines*, and *The Singularity is Near*.

My interest in the future stems from my interest in being an inventor. I've had the idea of being an inventor since I was five years old, and I quickly realized that you had to have a good idea of the future if you were going to succeed as an inventor. It's a little bit like surfing; you have to catch a wave at the right time. By the time you get something done, the world has become a different place from what it was when you started. Most inventors fail not because they can't get something to work but because the market's enabling forces are not all in place at the right time.

So I became a student of technology trends, and I have developed mathematical models of how technology evolves in various areas—like computers, electronics in general, communication storage devices, biological technologies like genetic scanning, reverse-engineering of the human brain, miniaturization, the size of technology, and the pace of paradigm shifts. This interest in trends took on a life of its own, and I began to project some of them using what I call the Law of Accelerating Returns, which I believe underlies the evolution of technology. A book I wrote in the 1980s, called *The Age of Intelligent Machines,* was a road map of what the 1990s and the early 2000s would be like, and it's worked out quite well. I've now refined those mathematical models and have begun to really examine what the twenty-first century will be like. It allows me to be inventive with the technologies of the twenty-first century, because I have a conception of what technology, communications, the size of technology, and our knowledge of the human brain will be like in 2010, 2020, and 2030. I can't actually create these technologies yet, but I can write about them. I've come to a view of the future that falls out of these models, which I believe are valid both for theoretical reasons and because they also match the empirical data of the twentieth century.

One thing that observers don't fully recognize, and that a lot of otherwise thoughtful people fail to take into consideration adequately, is that the pace of change itself has accelerated. Centuries ago people didn't think the world was changing at all. Their grandparents had the same lives that they did, and they expected their grandchildren would do the same, and that expectation was largely fulfilled. Today it's an axiom that life is changing and that technology is affecting the nature of society. But what's not fully understood is that the last twenty years are not a good guide to the next twenty years.

We're doubling the paradigm-shift rate, the rate of progress, every decade. This will actually match the amount of progress we made in the whole twentieth century, because we've been accelerating up to this point. The twentieth century was like twenty years' worth of change at today's rate of change. In the next twenty-five years we'll make three times the progress you saw in the twentieth century. And we'll make 20,000 years of progress in the twenty-first century, which is about 1,000 times more technical change than we saw in the twentieth century.

Specifically, computation is growing exponentially. The one exponential trend that people are aware of is Moore's Law. But Moore's Law is just one method for bringing exponential growth to computers. In accordance with Moore's Law, we can put twice as many transistors on an integrated circuit every twenty-four months. Being smaller, they also run faster, so this amounts to a quadrupling of computational power every twelve months. (The popular understanding of eighteen months is not correct and is not what Moore originally observed.)

What's not fully realized is that Moore's Law was not the first but the fifth paradigm to bring exponential growth to computers. We had electromechanical calculators, relay-based computers, vacuum tubes, and transistors. Every time one paradigm ran out of steam, another took over. For a while there were shrinking vacuum tubes, and finally they couldn't be made any smaller and still keep the vacuum, so transistors came along, as a whole different approach. There's been a lot of discussion about Moore's Law running out of steam in about twelve years, because by that time transistors will be only a few atoms in width and we won't be able to shrink them anymore—so that particular paradigm will run out of steam, too.

We'll then go to the sixth paradigm, which is massively parallel computing in three dimensions. We live in a three-dimensional world, and our brains are organized in three dimensions, so we might as well compute in three dimensions. The brain processes information using an electrochemical method 10 million times slower than electronics. But it makes up for this by being three-dimensional. Every interneuronal connection computes simultaneously, so you have 100 trillion things going on at the same time. And that's the direction we'll be going in. Right now, chips, even though they're very dense, are flat. Fifteen or twenty years from now, computers will be massively parallel and will be based on biologically inspired models, which we will devise largely by understanding how the brain works.

It's generally recognized that we'll have the hardware to recreate human intelligence within a brief period of time—I'd say about twenty years. What's more controversial is whether we will have the software. Observers acknowledge that we will have very fast computers that could in theory emulate the human brain, but we don't really know how the brain works, and we won't have the software, the methods, or the knowledge to create a human level of intelligence. Without this, you just have an extremely fast calculator.

But our knowledge of how the brain works is also growing exponentially. The brain is not of infinite complexity. It's a very complex entity, and we're not going to achieve a total understanding through one simple breakthrough, but we're further along in understanding the principles of operation of the human brain than most people realize. The technology for scanning the human brain is growing exponentially; our ability to actually see the internal connection patterns is growing, and we're developing more and more detailed mathematical models of biological neurons. We actually have very detailed mathematical models of several dozen regions of the human brain and how they work, and we've recreated their methodologies using conventional computation. The results of those reengineered or reimplemented synthetic models of those brain regions match the human brain very closely.

We're also replacing sections of the brain that are degraded or don't work anymore because of disabilities or disease. There are neural implants for Parkinson's disease and cochlear implants for deafness. There's a new generation of cochlear implants coming out that provide 1,000 points of frequency resolution and will allow deaf people to hear music for the first time. The Parkinson's implant replaces the cortical neurons destroyed by that disease. So we've shown that it's feasible to understand regions of the human brain and nervous system and reimplement those regions with conventional electronics computation that interacts with the brain and performs those functions.

If you follow these developments and work out the mathematics, it's a conservative scenario to say that within thirty years—possibly much sooner—we will have a complete map of the human brain, we will have complete mathematical models of how each region works, and we will be able to reimplement the methods of the human brain, which are quite different from many of the methods used in contemporary artificial intelligence. But they're similar to methods used in my own field, pattern recognition, which is the fundamental capability of the human brain. We can't think fast enough to logically analyze situations quickly, so we rely on our powers of pattern recognition. Within thirty years we'll be able to create nonbiological intelligence comparable to human intelligence.

Just as for a biological system, we'll have to provide it

with an education, but here we can bring to bear some of the advantages of machine intelligence: Once a machine has mastered particular skills, it can apply these skills much more quickly and accurately than unenhanced humans. A $1,000 computer can remember billions of things accurately—most of us are hard-pressed to remember a handful of phone numbers. Once they learn something, machines can also share their knowledge with other machines. We don't have quick downloading ports at the level of our interneuronal connection patterns and our concentrations of neurotransmitters, so we can't just download knowledge. I can't take my knowledge of French and download it to you, but machines can share their patterns of knowledge with each other. We can educate machines in a process that can be hundreds or thousands of times faster than the comparable process in humans. It can provide a twenty-year education to a human-level machine in perhaps a few weeks or a few days, and then those machines can share their knowledge.

The primary implication of all this will be to enhance our own human intelligence. We're going to be putting these machines inside our own brains. We're starting to do that now, in people who have severe medical problems and disabilities, but ultimately this will be happening to all of us. Without surgery, we'll be able to introduce nanoengineered machines into the bloodstream that can pass through the capillaries of the brain. These intelligent, blood-cell-size nanobots will actually be able to go to the brain and interact with biological neurons. The basic feasibility of communicating in both directions between electronic devices and biological neurons has already been demonstrated.

One application of sending billions of nanobots into the brain is full-immersion virtual reality. If you want to be in real reality, the nanobots sit there and do nothing, but if you want to go into virtual reality, the nanobots shut down the signals coming from your real senses and replace them with the signals you would be receiving if you were in the virtual environment. And you can go there with other people; you can have everything from sexual and sensual encounters to business negotiations in full-immersion virtual-reality environments that incorporate all of the senses. People will beam their own flow of sensory experiences and the neurological correlates of their emotions out into the Web, the way people now beam images from Webcams in their living rooms and bedrooms. This will enable you to plug in and actually experience what it's like to be someone else, including their emotional reactions, like the plot concept of *Being John Malkovich*. In virtual reality, you don't have to be the same person. You can be someone else; you can project yourself as a different person.

Most important, we'll be able to enhance our biological intelligence with nonbiological intelligence through intimate connections. This won't mean having just one thin pipe between the brain and a nonbiological system but actually having nonbiological intelligence in billions of different places in the brain. I don't know about you, but there are lots of books I'd like to read and Web sites I'd like to go to, and I find my mental bandwidth limiting. So instead of having a mere 100 trillion connections, we'll eventually have 100 trillion times a million. We'll be able to enhance our cognitive pattern-recognition capabilities greatly, think faster, and download knowledge.

If you follow these trends further, you get to a point where change is happening so rapidly that there appears to be a rupture in the fabric of human history. Some people have referred to this as the "Singularity." This is a term borrowed from physics, meaning a point of infinite density and energy that's a kind of rupture in the fabric of space-time. Here it's applied by analogy to human history, to the point where the rate of technological progress is so rapid that it appears as a rupture in the fabric of human history. In physics, it's impossible to see beyond a singularity, which creates an event boundary—and some people have hypothesized that it will be impossible to characterize human life after the Singularity. My question is, "What will human life will be like after the Singularity?" which I predict will occur somewhere right before the middle of the twenty-first century.

A lot of the concepts we have of the nature of human life—such as longevity—suggest a limited capability as biological, thinking entities. All of these concepts will undergo significant change as we basically merge with our technology. It's taken me a while to get my own mental arms around these issues. In *The Age of Intelligent Machines*, I ended with the specter of machines matching human intelligence somewhere between 2020 and 2050, and I basically have not changed my view on that time frame, although I've abandoned my view that this is a final specter. In the book I wrote ten years later, *The Age of Spiritual Machines*, I began to consider what life would be like past the point where machines could compete with us. Now I'm trying to consider what that will mean for human society.

One thing we should keep in mind is that innate biological intelligence is fixed. We have 10^{26} calculations per second in the human race (rounding to 10 billion human brains, each with about 100 billion neurons, with an average fan-out of 1,000 connections per neuron, each connection having a capacity of about 200 calculations per second). Fifty years from now, the biological intelligence of humanity will still be at the same order of magnitude. But machine intelligence is growing exponentially, and today it's a million times less than that biological figure. So although human intelligence is still dominant, the crossover point is around 2030, and nonbiological intelligence will continue its exponential rise.

This leads some people to ask how we can know whether another species or entity is more intelligent than we are. Isn't knowledge tautological? How can we know more than we do know? Who would know it, except us?

One response is not to want to be enhanced, not to have nanobots. A lot of people say that they just want to stay a biological person. But what will the Singularity look like to people who want to remain biological? The answer is that they won't really notice it, except for the fact that machine intelligence will appear to biological humanity to be their transcendent servants. These machines will appear very friendly, taking care of all of our needs. But providing that service of meeting all of the material and emotional needs of biological humanity will comprise a very tiny fraction of the mental output of the nonbiological component of our civilization. So there's a lot that biological humanity won't notice.

There are two levels of consideration here. On the economic level, mental output will be the primary criterion. We're already getting close to the point where the only thing that has value is information. Information has value to the degree that it reflects knowledge rather than just raw data. For example, a clock, a camera, a tape recorder are physical objects, but their real value is in the information that went into their design: the design of their chips and the software used to invent and manufacture them. The actual raw materials—a bunch of sand and some metals and so on—are worth a few pennies, but these products have value because of all the knowledge that went into creating them. And the knowledge component of products and services is asymptoting toward 100 percent. By the time we get to 2030, it will basically be 100 percent. Through a combination of nanotechnology and artificial intelligence, we'll be able to create virtually any physical product and meet all of our material needs. When everything is software and information, it will just be a matter of downloading the right software, and we're already getting pretty close to that.

On a spiritual level, the issue of what is consciousness is also important. We will have entities by 2030 that seem to be conscious and that will claim to have feelings. We have entities today—characters in your kids' video games, for instance—that can make comparable claims, but these claims aren't very convincing. They're software entities that are still a million times simpler than the human brain. In 2030, that won't be the case. Say you encounter another person in virtual reality who looks just like a human being but there's no biological human behind it—it's completely an AI projecting a humanlike figure in virtual reality, or even a humanlike image in real reality using an android robotic technology. These entities will seem human. They won't be a million times simpler than humans; they'll be as complex as humans. They'll have all the subtle cues of being humans. They'll be able to sit here and be interviewed and be just as convincing as a human, just as complex, just as interesting. And when they claim to be angry or happy, it will be just as convincing as when a human makes those claims.

At this point, we arrive at a deeply philosophical issue. Is such an entity just a very clever simulation good enough to trick you, or is it really conscious in the way we assume other people are? In my view, there's no real way to test that scientifically. There's no machine you can slide the entity into, in which a green light will go on and say, "OK, this entity's

conscious" or "This one's not." You could make such a machine, but it will have philosophical assumptions built into it.
Some philosophers will say that unless the entity has impulses squirting through biological neurotransmitters, it's
not conscious—or that unless it's a biological human with a
biological mother and father, it's not conscious. But consciousness becomes a matter of philosophical debate; it's not
scientifically resolvable.

The next big revolution—one that will affect us right
away—is biological technology, because we've merged biological knowledge with information processing. We are in
the early stages of understanding life processes and disease
processes by understanding the genome and how the
genome expresses itself in protein. And we're going to find—
and this has been apparent all along—that there's a slippery
slope and no clear definition of where life begins. Both sides
of the abortion debate have been afraid to get off the edges
of that debate: that life starts at conception or that life starts
at birth. They don't want to get off those edges, because they
realize it's just a completely slippery slope from one end to
the other. But we're going to make it even more slippery.
We'll be able to create stem cells without ever actually going
through the fertilized egg. What's the difference between a
skin cell, which contains the whole genome, and a fertilized
egg? The only differences are some proteins in the egg and
some signaling factors we don't yet fully understand, which
are basically proteins (small RNA molecules apparently play
a big role here). We will get to the point where we'll be able
to take some protein mix—which is just a bunch of chemicals
and clearly not a human being—and add it to a skin cell to
create a fertilized egg that we can then immediately differentiate into any cell of the body. When I rub my hands together
and brush off thousands of skin cells, I will be destroying

thousands of potential people. There won't be any clear boundary.

This is another way of saying that science and technology will find a way around the controversy. In the future, we'll be able to do therapeutic cloning, a very important technology, that completely avoids the concept of the fetus. We'll be able to take skin cells and create—pretty directly, without ever using a fetus—all the cells we need. There were significant advances in doing this just in the past year: Scientists were able to directly transform skin cells into immune system cells and nerve cells without using cloning or embryonic stem cells.

We're not that far away from being able to create new cells. For example, I'm fifty-four, but with my DNA I'll be able to create the heart cells of a twenty-five-year-old man, and I can replace my heart with those cells without surgery just by sending them through my bloodstream. They'll take up residence in the heart, so at first I'll have a heart that's 1 percent young cells and 99 percent older ones. But if I keep doing this every day, a year later my heart is 99 percent young cells. With that kind of therapy, we can ultimately replenish all the cell tissues and the organs in the body. This isn't something that will happen tomorrow, but these are the kinds of revolutionary processes we're on the verge of.

If you look at human longevity—another exponential trend—you'll notice that a few days were added every year to human life expectancy in the eighteenth century. In the nineteenth century, a few weeks were added every year, and now we're now adding over 100 days a year because of all these developments, which are going to continue to accelerate. Many knowledgeable observers, including myself, feel that within ten years we'll add more than a year per year to life expectancy. So as we get older, human life expectancy will expand at a faster rate than we're progressing in age. If we can hang in there, our generation is right on the edge. We have to watch our health the old-fashioned way for a while longer, so we're not the last generation to die prematurely. But by the time our children are thirty, forty years old, these technologies will be so advanced that human life expectancy will be vastly extended.

There is also the fundamental issue of whether ethical debates will stop the developments I'm talking about. It's all very well to have these mathematical models and trends, but the question is, Will they hit a wall because people for one reason or another—through war or ethical debates such as the stem-cell controversy—thwart this ongoing exponential development?

I strongly believe that won't happen. The ethical debates are like stones in a stream. The water runs around them. You haven't seen any biological technologies held up for one week by any of these debates. To some extent, we may have to find other ways around some of the limitations, but there are so many developments going on. There are dozens of very exciting ideas about how to use genomic information and proteomic information. Although the controversies may attach themselves to one idea here or there, there's such a river of advances—the very concept of technological advance is so deeply ingrained in our society—that it's an enormous imperative. I agree that the dangers are there, but it is not feasible to stop the accelerating progression of technology short of a *Brave New World* scenario of a totalitarian government using technology to ban all technology development.

The kinds of scenarios I'm talking about twenty or thirty years from now are not being developed because there's one laboratory sitting there creating a human-level intelligence in a machine. They are happening because that's the inevitable result of thousands of little steps. Each little step is conservative, not radical, and makes perfect sense. Each step is just the next generation of some company's products. If you take thousands of those little steps—which are happening faster and faster—you end up with remarkable changes ten, twenty, or thirty years from now. Despite the legitimate concerns of Sun Microsystems chief scientist Bill Joy, you don't hear Sun Microsystems saying that the future implication of these technologies is so dangerous that it's going to stop creating more intelligent networks and more powerful computers. Sun can't stop. No company can, because it would be out of business. There's an enormous economic imperative.

There is also a tremendous moral imperative. There still are not millions but billions of people who suffer from disease and poverty, and we have the opportunity to overcome those problems through technological advances. You can't tell the millions of people who are suffering from cancer that we're on the verge of great breakthroughs that will save them from cancer but we're stopping all that because terrorists might use that same knowledge to create a bioengineered pathogen. Such expropriation is a valid concern, but we're not going to stop. In our society there's a tremendous belief in the benefits of continued economic and technological advance. Still, it does raise the question of the dangers of these technologies. I do agree that we need to focus our attention on dealing specifically with these scenarios of danger. It is, in my view, the primary challenge of the twenty-first century.

Another aspect of all these changes is that they force us to reevaluate our concept of what it means to be human. There is a common objection to the advance of technology and its implications for humanity. It goes like this: We'll have very powerful computers, but we haven't solved the software problem, and because the software is so incredibly complex, we can't manage it. I address this objection by saying that the software required to emulate human intelligence is actually not beyond our current capability. We have to use different techniques—different self-organizing methods—that are biologically inspired. The brain is complicated, but it's not that complicated. You have to keep in mind that it is characterized by a genome of only 23 million bytes. The genome is 6 billion bits—that's 800 million bytes—and there are massive redundancies. One pretty long sequence called ALU is repeated 300,000 times. If you use conventional data compression on the genome, you get about 23 million bytes (a small fraction of the size of Microsoft Word), which is a level of complexity we can handle. But we have not yet reverse-engineered this information—that is, we don't yet understand the principles of operation of the human brain.

You might wonder how something with 23 million bytes can create a human brain that's a million times more complicated than itself. That's not hard to understand. The genome creates a process of wiring a region of the human brain involving a lot of randomness. Then, when the fetus becomes a baby and interacts with a very complicated world, there's an evolutionary process within the brain in which a lot of the connections die out, others get reinforced, and it self-organizes to represent meaningful knowledge and skills. It's a very clever system, and we don't understand it yet, but we will, because it's not a level of complexity beyond what we're capable of engineering.

In my view, there is something special about human beings that's different from what we see in any of the other animals. By happenstance of evolution, we were the first species to be able to create technology. Actually there were others, but we are the only one that survived in this ecological niche. We combined a rational faculty, the ability to think logically, to create abstractions, to create models of the world in our own minds, and to manipulate the world. We have opposable thumbs, so that we can create technology, but technology is not just tools. Other animals have used primitive tools. The difference is a body of knowledge that changes and evolves from generation to generation. The knowledge that the human species has is another one of those exponential trends.

We use one stage of technology to create the next stage, which is why technology accelerates, why it grows in power. Today, for example, a computer designer has tremendously powerful computer-system-design tools to create computers, so in a couple of days they can create a very complex system and it can all be worked out very quickly. The first computer designers had to actually draw them all out in pen on paper. Each generation of tools creates the power to create the next generation.

So technology itself is an exponential, evolutionary process, a continuation of the biological evolution that created humanity in the first place. Biological evolution itself evolved in an exponential manner. Each stage created more powerful tools for the next, so when biological evolution created DNA it now had a means of keeping records of its experiments so that evolution could proceed more quickly. Because of this, the Cambrian explosion lasted only a few tens of millions of years, whereas the first stage—that of creating DNA and primitive cells—took billions of years. Finally, biological evolution created a species that could manipulate its environment and had some rational faculties, and now the cutting edge of evolution has changed from biological evolution into something carried out by one of its own creations, *Homo sapiens*, and is represented by technology. In the next epoch, this species that ushered in its own evolutionary process—that is, its own cultural and technological evolu-tion, as no other species has—will combine with its own cre-ation. It will merge with its technology. At some level that's already happening—even if most of us don't necessarily have it yet inside our bodies and brains—since we're very intimate with technology. It's in our pockets.

ONE HALF OF
A MANIFESTO

Jaron Lanier

We imagine "pure" cybernetic systems, but we can prove only that we know how to build fairly dysfunctional ones. We kid ourselves when we think we understand something, even a computer, merely because we can model or digitize it.

JARON LANIER, a computer scientist and musician, is lead scientist of the National Tele-Immersion Initiative, a coalition of research universities studying advanced applications for Internet 2. Best known for his work in virtual reality, a term he coined, Lanier helped develop the first implementations of multiperson virtual worlds using head-mounted displays. He also codeveloped the first implementations of virtual reality in surgical simulation, vehicle design prototyping, and various other applications. As a musician, he writes for orchestra, plays a large number of instruments from around the world, and has performed with a wide variety of collaborators, from Philip Glass to George Clinton.

For the last twenty years, I have found myself on the inside of a revolution but on the outside of its resplendent dogma. Now that the revolution has not only hit the mainstream but bludgeoned it into submission by taking over the economy, it's probably time for me to cry out my dissent more loudly than I have before.

The dogma I object to is composed of a set of interlocking beliefs and doesn't have a generally accepted overarching name as yet, though I sometimes call it "cybernetic totalism." It has the potential to transform human experience more powerfully than any prior ideology, religion, or political system, partly because it can be so pleasing to the mind (at least initially) but mostly because it gets a free ride on the powerful technologies created by people who are, to a large degree, true believers.

The original usage of the term "cybernetic," as by Norbert Weiner, was certainly not restricted to digital computers. It was originally meant to suggest a metaphor between marine navigation and a feedback device that governs a mechanical system, such as a thermostat. Weiner certainly recognized and humanely explored the extraordinary reach of this metaphor, one of the most powerful ever expressed. I hope no one will think I'm equating cybernetics and what I'm calling cybernetic totalism. The distance between recognizing a great metaphor and treating it as the only metaphor is the same as the distance between humble science and dogmatic religion.

Here is a partial roster of the component beliefs of cybernetic totalism:

1) That cybernetic patterns of information provide the ultimate and best way to understand reality.
2) That people are no more than cybernetic patterns.
3) That subjective experience either doesn't exist or is unimportant because it is some sort of ambient or peripheral effect.
4) That what Darwin described in biology, or something like it, is in fact the singular, superior description of all creativity and culture.
5) That qualitative as well as quantitative aspects of information systems will be accelerated by Moore's Law.

And finally, the most dramatic:

6) That biology and physics will merge with computer science, becoming biotechnology and nanotechnology,

resulting in life and the physical universe becoming mercurial—and achieving the supposed nature of computer software. Furthermore, all of this will happen very soon! Since computers are improving so quickly, they will overwhelm all the other cybernetic processes (like people) and fundamentally change the nature of what's going on in the familiar neighborhood of Earth at some moment when a new "criticality" is achieved—maybe in about the year 2020. To be a human after that moment will be either impossible or else something very different than we now can know.

During the last twenty years, a stream of books has gradually informed the larger public about the belief structure of the inner circle of digerati, starting softly—for instance, with Douglas Hofstadter's *Gödel, Escher, Bach*—and growing more harsh with recent entries, such as Ray Kurzweil's *The Age of Spiritual Machines.*

Recently, public attention has finally been drawn to #6—the astonishing belief in an eschatological cataclysm in our lifetimes, brought about when computers become the ultraintelligent masters of physical matter and life. As far as I can tell, a large number of my friends and colleagues believe in some version of this imminent doom. I am curious about who among the celebrated thinkers who largely accept some version of the first five points are also comfortable with the sixth, the eschatology. In general, I find that technologists rather than natural scientists have tended to be vocal about the possibility of a near-term criticality. I have no idea, however, what figures like the evolutionary biologist Richard Dawkins or the philosopher Daniel Dennett make of it. Somehow I can't imagine these elegant theorists speculating about whether nanorobots might take over the planet in twenty years. It seems beneath their dignity. Yet the eschatologies of Kurzweil, Hans Moravec, and Eric Drexler follow directly—and it would seem inevitably—from an understanding of the world that has been most sharply articulated by none other than Dawkins and Dennett. Do Dawkins, Dennett, and others in their camp see some flaw in logic that insulates their thinking from the eschatological implications? The primary candidate for such a flaw, as I see it, is that cyber-armageddonists have confused ideal computers with real computers, which behave differently. My position on this point can be evaluated separately from my admittedly provocative positions on the first five points, and I hope it will be.

Why is this only "One Half of a Manifesto"? I hope readers will not think I've sunk into some sort of glum rejection of digital technology. In fact, I'm more delighted than ever to be working in computer science, and I find that it's rather easy to adopt a humanistic framework for designing digital tools. There is a lovely global flowering of computer culture already in place, arising for the most part independently of the technological elites, which implicitly rejects the ideas I am attacking here. A full manifesto would attempt to describe and promote this positive culture.

I will now examine the five beliefs that must precede acceptance of the new eschatology, and then consider the eschatology itself.

- *Cybernetic Totalist Belief #1: That cybernetic patterns of information provide the ultimate and best way to understand reality.* There is an undeniable rush of excitement experienced by those who first are able to perceive a phenomenon cybernetically. For example, while I believe I can imagine what a thrill it must have been to use early photographic equipment in the nineteenth century, I can't imagine that any outsider could comprehend the sensation of being around early computer graphics technology in the 1970s. For here was not merely a way to make and show images but a meta-framework that subsumed all possible images. Once you can understand something in a way that you can shove it into a computer, you have cracked its code, transcended any particularity it might have at a given time. It was as if we had become the gods of vision and had effectively created all possible images, for they would merely be reshufflings of the bits in the computers we had before us, completely under our control.

The cybernetic impulse is initially driven by ego (though, as we shall see, in its endgame, which has not yet arrived, it will become the enemy of ego). For instance, cybernetic totalists look at culture and see "memes"—autonomous mental tropes that compete, somewhat like viruses, for brain space in humans. In doing so, the cybernetic totalists not only accomplish a triumph of campus imperialism, placing themselves in an imagined position of superior understanding versus the whole of the humanities, but they also avoid having to pay much attention to the particulars of culture in a given time and place. Once you have subsumed something into its cybernetic reduction, any particular reshuffling of its bits seems unimportant.

Belief #1 appeared almost immediately with the first computers. It was articulated by the first generation of computer scientists—Weiner, Shannon, Turing. It is so fundamental that it isn't even stated anymore within the inner circle. It is so well rooted that it is difficult for me to remove myself from my all-encompassing intellectual environment long enough to articulate an alternative to it. But an alternative might be this: A cybernetic model of a phenomenon can never be the sole favored model, because we can't even build computers that conform to such models. Real computers are completely different from the ideal computers of theory. They break for reasons that are not always analyzable and they seem to intrinsically resist many of our endeavors to improve them—in large part due to legacy and lock-in, among other problems. We imagine "pure" cybernetic systems, but we can prove only that we know how to build fairly dysfunctional ones. We kid ourselves when we think we understand something, even a computer, merely because we can model or digitize it.

There is also an epistemological problem that bothers me, even though my colleagues by and large are willing to ignore it. I don't think you can measure the function or even the existence of a computer without a cultural context for it. I don't think Martians would necessarily be able to distinguish a Macintosh from a space heater.

The above disputes ultimately turn on a combination of technical arguments about information theory and philosophical positions that largely arise from taste and faith. So I try to augment my positions with pragmatic considerations, and some of these will begin to appear in my thoughts on . . .

- *Belief #2: That people are no more than cybernetic patterns.* Every cybernetic totalist fantasy relies on artificial intelligence. It might not immediately be apparent why such fantasies are essential to those who have them. If computers are to become smart enough to design their own successors, initiating a process that will lead to godlike omniscience after a number of ever-swifter passages from one generation of computers to the next, someone is going to have to write the software that gets the process going, and humans have given absolutely no evidence of being able to write such software. So the idea is that the computers will somehow become smart on their own and write their own software.

My primary objection to this way of thinking is pragmatic: It results in the creation of poor-quality real-world software in the present. Cybernetic totalists live with their heads in the future and are willing to accept obvious flaws in present software in support of a fantasy world that might never appear.

The whole enterprise of artificial intelligence is based on an intellectual mistake and continues to expensively turn out poorly designed software remarketed under a new name for every new generation of programmers. Lately it has been called "intelligent agents"; the last time around it was called "expert systems."

Let's start at the beginning, when the idea first appeared. In Turing's famous thought experiment, a human judge is asked to determine which of two correspondents is human and which is machine. If the judge cannot tell, Turing asserts that the computer should be treated as having essentially achieved the moral and intellectual status of personhood. Turing's mistake was that he assumed that the only explanation for a successful computer entrant would be that the computer had become elevated in some way, by becoming smarter, more human. There is another, equally valid explanation of a winning computer, however, which is that the human had become less intelligent, less humanlike. An official Turing test is held every year, and while the substantial cash prize has not been claimed by a program yet, it will certainly be won sometime in the coming years. My view is that this event is distracting everyone from the real Turing tests that are already being won. Real, though miniature, Turing tests are happening all the time, every day, whenever a person puts up with stupid computer software.

For instance, in the United States we organize our financial lives in order to look good to the pathetically simplistic computer programs that determine our credit ratings. We borrow money when we don't need to, for example, to feed the type of data to the programs that we know they are programmed to respond to favorably. In doing this, we make ourselves stupid in order to make the computer software seem smart. In fact, we continue to trust the credit-rating software even though there has been an epidemic of personal bankruptcies during a time of very low unemployment and great prosperity. We have caused the Turing test to be passed. There is no epistemological difference between artificial intelligence and the acceptance of badly designed computer software.

My argument can be taken as an attack against the belief in eventual computer sentience, but a more sophisticated reading would be that it argues for a pragmatic advantage to holding an anti-AI belief, since those who believe in AI are more likely to put up with bad software. More important, I'm hoping the reader can see that artificial intelligence is better understood as a belief system than as a technology.

- *Belief #3: That subjective experience either doesn't exist or is unimportant because it is some sort of ambient or peripheral effect.* There is a new moral struggle taking shape over the question of when "souls" should be attributed to perceived patterns in the world. Computers, genes, and the economy are some of the entities that cybernetic totalists think of as populating reality today along with human beings. It is certainly true that we are confronted with nonhuman and metahuman actors in our lives on a constant basis, and these players sometimes appear to be more powerful than us. So the new moral question is, Do we make decisions solely on the basis of the needs and wants of "traditional" biological humans or are any of these other players deserving of consideration?

I propose to make use of a simple image to consider the alternative points of view. This image is of an imaginary circle that each person draws around him/herself. We shall call this "the circle of empathy." On the inside of the circle are those things that are considered deserving of empathy and the corresponding respect, rights, and practical treatment as approximate equals. On the outside of the circle are those things that are considered less important, less alive, less deserving of rights. (This image is only a tool for thought, and should certainly not be taken as my complete model for human psychology or moral dilemmas.) Roughly speaking, liberals hope to expand the circle, while conservatives wish to contract it.

Should computers, perhaps at some point in the future, be placed inside the circle of empathy? The idea that they should is held close to the heart by the cybernetic totalists, who populate the elite technological academies and the businesses of the "new economy."

There has often been a tender but unintended humor in the argumentative writing by advocates of eventual computer sentience. The quest to rationally prove the possibility of sentience in a computer (or perhaps in the Internet) is the modern version of proving God's existence. As is the case with the history of God, a great many great minds have spent excesses of energy on this quest, and eventually a cybernetically minded twenty-first-century version of Kant will appear in order to present a tedious "proof" that such adventures are futile. I simply don't have the patience to be that person. As it happens, in the last five years or so, arguments about computer sentience have started to subside. The idea is assumed to be true by most of my colleagues; for them, the argument is over. It is not over for me.

I must report that back when those arguments were still white-hot, it was the oddest feeling to debate someone like cybernetic totalist philosopher Daniel Dennett. He would state that humans were simply specialized computers, and that imposing some fundamental ontological distinction between humans and computers was a sentimental waste of time. "But don't you experience your life?" I would ask. "Isn't experience something apart from what you could measure in a computer?" My opponent would typically counter with something like, "Experience is just an illusion created because there is one part of a machine [you] that needs to create a model of the function of the rest of the machine—that part is your experiential center." I would retort that experience is the only thing that isn't reduced by illusion—that even illusion itself is experience. A correlate, alas, is that experience is the very thing that can only be experienced. This led me into the odd position of publicly wondering if some of my opponents simply lacked internal experience. (I once suggested that among all humanity, one could definitively prove a lack of internal experience only in certain professional philosophers.) In truth, I think my perennial antagonists do have internal experience but choose not to admit it in public, for a variety of reasons, most often because they enjoy annoying people.

Another motivation might be the "campus imperialism" mentioned earlier. Representatives of each academic discipline occasionally assert that they have a privileged viewpoint that somehow contains or subsumes the viewpoints of their rivals. Physicists were the alpha academics for much of the twentieth century, though in recent decades "postmodern" humanities thinkers managed to stage something of a comeback, at least in their own minds. But technologists are the inevitable winners of this game, as they pull the very components of our lives out from under us. It is tempting to many of them, apparently, to bolster this power by suggesting that they also possess an ultimate understanding of reality, which is something quite apart from having tremendous influence on it.

A third motivation might be neo-Freudian, considering that the primary advocate of the idea of machine sentience, Alan Turing, was such a tortured soul. Turing died in an apparent suicide brought on by his having developed breasts as a result of enduring a hormonal regimen intended to reverse his homosexuality. It was during this tragic final period of his life that he argued passionately for machine sentience, and I have wondered whether he was engaging in a highly original form of psychological escape and denial—running away from sexuality and mortality by becoming a computer.

At any rate, what is peculiar and revealing is that my cybernetic totalist friends confuse the viability of a perspective with its triumphant superiority. It is perfectly true that one can think of a person as a gene's way of propagating itself, as per Dawkins, or as a sexual organ used by machines to make more machines, as per Marshall McLuhan (as quoted in the masthead of every issue of *Wired*), and indeed it can even be beautiful to think from these perspectives from time to time. As the anthropologist Steve Barnett has pointed out, however, it would be just as reasonable to assert that "A person is shit's way of making more shit."

So let us pretend that the new Kant has already appeared and done his/her inevitable work. We can then say: The placement of one's circle of empathy is ultimately a matter of faith. We must accept the fact that we are forced to place the circle somewhere, and yet we cannot exclude extrarational faith from our choice of where to place it. My personal choice is to not place computers inside the circle. In this essay, I am stating some of my pragmatic, aesthetic, and political reasons for this choice, though ultimately my decision rests on my particular faith.

- *Belief #4: That what Darwin described in biology, or something like it, is in fact also the singular, superior description of all possible creativity and culture.* Cybernetic totalists are obsessed with Darwin, for he described the closest thing we have to an algorithm for creativity. Darwin answers what would otherwise be a big hole in the Dogma: How will cybernetic systems be smart and creative enough to invent a posthuman world? In order to embrace an eschatology in which the computers become smart as they become fast, some kind of *deus ex machina* must be invoked, and it has a beard.

Unfortunately, in the current climate I must take a moment to state that I am not a creationist. I am in this essay criticizing what I perceive to be intellectual laziness—a retreat from trying to understand problems and instead hoping for software that evolves itself. I am *not* suggesting that nature required some extra element beyond natural evolution to create people. I also don't mean to imply that there is a completely unified block of people opposing me, all of whom think exactly the same thoughts. There are, in fact, numerous variations of Darwinian eschatology. Some of the most dramatic renditions have come not from scientists or engineers but from writers such as Kevin Kelly and Robert Wright, who have become entranced with broadened interpretations of Darwin. In their works, reality is perceived as a big computer program running the Darwin algorithm, perhaps headed toward some sort of Destiny.

Many of my technical colleagues also see at least some form of a causal arrow in evolution pointing to an ever-greater degree of a hard-to-characterize something as time passes. The words used to describe that something are themselves hard to define; it is said to include increased complexity, organization, and representation. To the computer scientist Danny Hillis, people seem to have more of such a thing than, say, single-cell organisms, and it is natural to wonder if perhaps there will someday be some new creatures with even more of it than is found in people. (And, of course, the future birth of the new "more so" species is usually said to be related to computers.) Contrast this perspective with that of Stephen Jay Gould, who argued in *Full House* that if there's an arrow in evolution, it's toward greater diversity over time, and we unlikely creatures known as humans, having arisen as one tiny manifestation of a massive, blind exploration of possible creatures, only imagine that the whole process was designed to lead to us.

There is no harder idea to test than an anthropic one, or its refutation. I'll admit that I tend to side with Gould on this one, but it is more important to point out an epistemological conundrum that should be considered by Darwinian eschatologists. If humankind is the measure of evolution thus far, then we will also be the measure of successor species that might be purported to be "more evolved" than us. We'll have to anthropomorphize in order to perceive this "greater than human" form of life, especially if it exists inside an information space such as the Internet.

In other words, we'll be as reliable in assessing the status of the new superbeings as we are in assessing the traits of pet dogs in the present. We aren't up to the task. Before you tell me that it will be overwhelmingly obvious when the superintelligent new cyberspecies arrives, visit a dog show. Or a gathering of people who believe they have been abducted by aliens in UFOs. People are demonstrably insane when it comes to assessing nonhuman sentience.

There is, however, no question that the movement to interpret Darwin more broadly, and in particular to bring him into psychology and the humanities, has offered some luminous insights that will someday be part of an improved understanding of nature, including human nature. I enjoy this stream of thought on various levels. It's also, let's admit it, impossible for a computer scientist not to be flattered by works that place what is essentially a form of algorithmic computation at the center of reality, and these thinkers tend to be confident and crisp and to occasionally have new and good ideas.

And yet I think cybernetic totalist Darwinians are often incompetent at public discourse and may be in part responsible, however unintentionally, for inciting a resurgence of fundamentalist religious reaction against rational biology. They seem to come up with takes on Darwin that are calculated not only to antagonize but to alienate those who don't share their views. Declarations from the nerdiest of the evolutionary psychologists can be particularly irritating. One example is the recent book by Randy Thornhill and Craig T. Palmer called *The Natural History of Rape*, which declares that rape is a "natural" way to spread genes around. We have seen all sorts of propositions tied to Darwin with a veneer of rationality. In fact, you can argue almost any position using a Darwinian strategy. For instance, Thornhill and Palmer suggest that those who disagree with them are victims of evolutionary programming for the need to believe in a fictitious altruism in human nature. The authors say it is seemingly altruistic to disbelieve in evolutionary psychology, because such skepticism makes a public display of one's belief in brotherly love. Displays of altruism are said to be attractive and therefore to improve one's ability to lure mates. By this logic, evolutionary psychologists should soon breed themselves out of the population. Unless they resort to rape.

At any rate, Darwin's idea of evolution was of a different order than previous scientific theories, for at least two reasons. The most obvious and explosive reason was that the subject matter was so close to home. It was a shock to the nineteenth-century mind to think of animals as blood relatives, and that shock continues to this day. The second reason is less often recognized. Darwin created a style of reduction that was based on emergent principles instead of underlying laws. There isn't any evolutionary "force" analogous to, say, electromagnetism. Evolution is a principle that can be discerned as emerging in events, but it cannot be described precisely as a force that directs events. This is a subtle distinction. The story of each photon is the same, in a way that the story of each animal and plant is different. (Of course, there are wonderful examples of precise, quantitative statements in Darwinian theory and corresponding experiments, but these don't take place at anywhere close to the level of human experience, which is whole organisms that have complex behaviors in environments.) "Story" is the operative word. Evolutionary thought has almost always been applied to specific situations through stories. A story, unlike a theory, invites embroidery and variation; indeed, stories gain their communicative power by resonance with more primal stories. It is possible to learn physics without inventing a narrative in one's head to give meaning to photons and black holes. But it seems that it is impossible to learn Darwinian evolution without also developing an internal narrative to relate it to other stories one knows. At least no public thinker on the subject seems to have confronted Darwin without building a bridge to personal value systems.

But beyond the question of subjective flavoring, there remains the problem of whether Darwin has explained enough. Is it not possible that there remains an as-yet-unarticulated idea that explains aspects of achievement and creativity that Darwin does not? For instance, is Darwinian-styled explanation sufficient to understand the process of rational thought? There are a plethora of recent theories in which the brain is said to produce random distributions of subconscious ideas that compete with one another until only the best one has survived, but do these theories really fit with what people do? In nature, evolution appears to be brilliant at optimizing but stupid at strategizing. (The mathematical image that expresses this idea is that "blind" evolution has enormous trouble getting unstuck from local minima in an energy landscape.) The classic question would be, How could evolution have made such marvelous feet, claws, fins, and paws but have missed the wheel? There are plenty of environments in which creatures would benefit from wheels, so why haven't any appeared? Not even once? (A great long-term art project for some rebellious kid in school now: Genetically engineer an animal with wheels! See if DNA can be made to do it.)

People came up with the wheel and numerous other use-ful inventions that seem to have eluded evolution. It is possi-ble that the explanation is simply that hands had access to a different set of inventions than DNA did, even though both were guided by similar processes. But it seems to me premature to treat such an interpretation as a certainty. Is it not possible that in rational thought the brain does some as-yet-unarticulated thing that might have originated in a Darwinian process but that cannot be explained by it?

The first two or three generations of artificial-intelligence researchers took it as a given that blind evolution in itself couldn't be the whole of the story and assumed that there were elements that distinguished human mentation from other earthly processes. For instance, humans were thought by many to build abstract representations of the world in their minds, while the process of evolution needn't do that. Furthermore, these representations seemed to possess extraordinary qualities, like the fearsome and perpetually elusive "common sense." After decades of failed attempts to build similar abstractions in computers, the field of AI gave up, but without admitting it. Surrender was couched as merely a series of tactical retreats. AI these days is often conceived as more of a craft than a branch of science or engineering. A great many practitioners I've spoken with lately hope to see software evolve that does various things, but they seem to have sunk to an almost postmodern, or cynical, lack of concern with understanding how these gizmos might actually work.

It is important to remember that craft-based cultures can come up with plenty of useful technologies, and that the motivation for our predecessors to embrace the Enlightenment and the ascent of rationality was not just to make more technologies more quickly. There was also the idea of humanism and a belief in the goodness of rational thinking and understanding. Are we really ready to abandon that?

Finally, there is an empirical point to be made: There has now been over a decade of work worldwide in Darwinian approaches to generating software, and while there have been some fascinating and impressive isolated results (and indeed I enjoy participating in such research), nothing has arisen from the work that would make software in general any better. So, while I love Darwin, I wouldn't count on him to write code.

- *Belief #5: That qualitative as well as quantitative aspects of information systems will be accelerated by Moore's Law.* The hardware side of computers keeps on getting better and cheaper at an exponential rate known as Moore's Law: Every year and a half or so, computation gets roughly twice as fast for a given cost. The implications of this are dizzying and so profound that they induce vertigo on first apprehension. What could a computer that was a *million* times faster than the one I am writing this text on be able to do? Would such a computer really be incapable of doing whatever it is that my human brain does? The quantity of a "million" is not only too large to grasp intuitively, it is not even accessible experimentally for present purposes, so speculation is not irrational. What is stunning is to realize that many of us will find out the answer in our lifetimes, for such a computer might be a cheap consumer product in about, say, thirty years.

This breathtaking vista must be starkly contrasted with the Great Shame of computer science, which is that we don't seem to be able to write software much better as computers get much faster. Computer software continues to disappoint. How I hated Unix back in the seventies—that devilish accumulator of data trash, obscurer of function, enemy of the user! If anyone had told me then that getting back to the embarrassingly primitive Unix would be the great hope and investment obsession of the year 2000, merely because its name was changed to Linux and its source code was opened up again, I never would have had the stomach or the heart to continue in computer science.

If anything, there's a reverse Moore's Law observable in software: As processors become faster and memory becomes cheaper, software becomes correspondingly slower and more bloated, using up all available resources. Now, I know I'm not being entirely fair here. We have better speech recognition and language translation than we used to, for example, and we are learning to run larger databases and networks. But our core techniques and technologies for software simply haven't kept up with hardware. (Just as some newborn race of super-intelligent robots are about to consume all humanity, our dear old species will likely be saved by a Windows crash. The poor robots will linger pathetically, begging us to reboot them, even though they will know that rebooting will do no good.)

There are various reasons that software tends to be un-wieldy, but a primary one is what I like to call "brittleness." Software breaks before it bends, so it demands perfection in a universe that prefers statistics. This in turn leads to all the pain of legacy/lock-in and other perversions. The distance between the ideal computers we imagine in our thought experiments and the real computers we know how to unleash on the world could not be more discouraging.

It is the fetishizing of Moore's Law that seduces researchers into complacency. If you have an exponential force on your side, surely it will ace all challenges. Who cares about rational understanding, when you can instead rely on an exponential extrahuman fetish? But processing power isn't the only thing that scales impressively; so do the problems that processors have to solve.

Here's an example I offer to nontechnical people to illustrate this point. Ten years ago, I had a laptop with an indexing program that let me search for files by content. In order to respond quickly enough when I performed a search, it went through all the files in advance and indexed them, just as search engines like Google index the Internet today. The indexing process took about an hour. Today I have a laptop that is hugely more capacious and faster in every dimension, as predicted by Moore's Law. However, I now have to let my indexing program run overnight to do its job. There are many other examples of computers seeming to get slower even though central processors are getting faster. Computer user interfaces tend to respond more slowly to user interface events, such as a keypress, than they did fifteen years ago, for instance. What's gone wrong?

The answer is complicated.

One part of the answer is fundamental. It turns out that when programs and datasets get bigger (and increasing storage and transmission capacities are driven by the same processes that drive Moore's exponential speed-up), internal computational overhead often increases at a worse-than-linear rate. This is because of some nasty mathematical facts of life regarding algorithms. Making a problem twice as large usually makes it take a lot more than twice as long to solve. Some algorithms are worse in this way than others, and one aspect of a solid undergraduate education in computer science is learning about them. Plenty of problems have overheads that scale even more steeply than Moore's Law. Surprisingly few of the most essential algorithms have overheads that scale at a merely linear rate.

But that's only the beginning of the story. It's also true that if different parts of a system scale at different rates (and that's usually the case), one part might be overwhelmed by the other. In the case of my indexing program, the size of hard disks actually grew faster than the speed of interfaces to them. Overhead costs can be amplified by such examples of "messy" scaling, in which one part of a system cannot keep up with another. A bottleneck then appears, rather like grid-lock in a poorly designed roadway—and the backup that re-sults is just as bad as a morning commute on a typically inadequate roadway system. And just as tricky and expensive to plan for and prevent. (Trips on Manhattan streets were faster 100 years ago than they are today. Horses are faster than cars.)

And *then* we come to our old antagonist, brittleness. The larger a piece of computer software gets, the more it is likely to be dominated by some form of legacy code and the more brutal becomes the overhead of addressing the endless exam-ples of subtle incompatibility that inevitably arise between chunks of software originally created in different contexts. And even beyond these effects, there are failings of human character that worsen the state of software, and many of these are systemic and might arise even if nonhuman agents were writing the code. For instance, it is very time-consuming and expensive to plan ahead to make the tasks of future program-mers easier, so each programmer tends to choose strategies that worsen the effects of brittleness. The time crunch faced by programmers is driven by none other than Moore's Law, which motivates an ever faster turnaround of software revi-sions to get at least some form of mileage out of increasing processor speeds. So the result is often software that becomes less efficient in some ways even as processors become faster.

I see no evidence that Moore's Law is steep enough to outrun all these problems without additional unforeseen in-tellectual achievements.

A fundamental statement of the question I'm examining here is, Does software tend to be unwieldy only because of human error, or is the difficulty intrinsic to the nature of software itself? If there is any credibility at all to the eschatological scenarios of Kurzweil, Drexler, Moravec, et al., then this is the single most important question related to the future of humankind.

There is at least some metaphorical support for the possibility that software unwieldiness is intrinsic. In order to examine this possibility, I'll have to break my own rule and be a cybernetic totalist for a moment. Nature might seem less brittle than digital software, but if species are thought of as "programs," then it looks as if nature also has a software crisis. Evolution itself has evolved—introducing sex, for instance—but evolution has never found a way to be any speed but very slow. This might be at least in part because it takes a long time to explore the space of possible variations of an exceedingly vast and complex causal system to find new configurations that are viable. Natural evolution's slowness as a medium of transformation is apparently systemic, rather than resulting from some inherent sluggishness in its component parts. On the contrary, adaptation is capable of achieving thrilling speed, in select circumstances. An example of fast change is the adaptation of germs to our efforts to eradicate them. Resistance to antibiotics is a notorious contemporary example of biological speed.

Both human-created software and natural selection seem to accrue hierarchies of layers that vary in their potential for speedy change. Slow-changing layers protect local theaters within which there is a potential for faster change. In computers, this is the divide between operating systems and applications, or between browsers and Web pages. In biology it might be seen, for example, in the divide between nature- and nurture-dominated dynamics in the human mind. But the lugubrious layers seem to usually define the overall character and potential of a system.

In the minds of some of my colleagues, all you have to do is identify one layer in a cybernetic system that's capable of fast change and then wait for Moore's Law to work its magic. For instance, even if you're stuck with Linux, you might implement a neural net program in it that eventually grows huge and fast enough (because of Moore's Law) to achieve a moment of insight and rewrite its own operating system. The problem is that in every example we know, a layer that can change fast also can't change very much. Germs can adapt to new drugs quickly but would still take a very long time to evolve into owls. This might be an inherent trade-off. For an example in the digital world, you can write a new Java applet pretty quickly, but it won't look much different from other quickly written applets; take a look at what's been done with applets and you'll see that this is true.

Now we finally come to . . .

- *Belief #6, the coming cybernetic cataclysm.* When a thoughtful person marvels at Moore's Law, there might be awe and there might be terror. One version of this terror was expressed recently by Bill Joy, in a cover story for *Wired*. Bill accepts the pronouncements of Ray Kurzweil and others who believe that Moore's Law will lead to autonomous machines, perhaps by the year 2020. That is when computers will become, according to some estimates, about as powerful as human brains. (Not that anyone knows enough to really measure brains against computers yet. But for the sake of argument, let's suppose that the comparison is meaningful.) According to this scenario of the Terror, computers won't be stuck in boxes. They'll be more like robots, all connected together on the Net, and they'll have quite a bag of tricks.

They'll be able to perform nanomanufacturing, for one thing. They'll quickly learn to reproduce and improve themselves. One fine day, without warning, the new supermachines will brush humanity aside as casually as humans clear a forest for a new development. Or perhaps the machines will keep humans around to suffer the sort of indignity portrayed in the movie *The Matrix*. Even if the machines would otherwise choose to preserve their human progenitors, evil humans will be able to manipulate the machines to do vast harm to the rest of us. This is a different scenario, which Bill also explores. Biotechnology will have advanced to the point that computer programs will be able to manipulate DNA as if it were Javascript. If computers can calculate the effects of drugs, genetic modifications, and other biological trickery, and if the tools to realize such tricks are cheap, then all it takes is one madman to, say, create an epidemic targeting a single race. Biotechnology without a strong, cheap information technology component would not be sufficiently potent to bring about this scenario. Rather, it is the ability of software running on fabulously fast computers to cheaply model and guide the manipulation of biology that is at the root of this variant of the Terror. I haven't been able to fully convey Bill's concerns in this brief account, but you get the idea.

My version of the Terror is different. We can already see how the biotechnology industry is setting itself up for decades of expensive software trouble. While there are all sorts of useful databases and modeling packages being developed by biotech firms and labs, they all exist in isolated developmental bubbles. Each such tool expects the world to conform to its requirements. Since these tools are so valuable, the world will do exactly that, but we should expect to see vast resources applied to the problem of getting data from one bubble into another. There is no giant monolithic electronic brain being created with biological knowledge. There is instead a fractured mess of data and modeling fiefdoms. The medium for biological data transfer will continue to be sleep-deprived individual human researchers until some fabled future time when we know how to make software that is good at bridging bubbles on its own.

What is a long-term future scenario like in which hardware keeps getting better and software remains mediocre? The great thing about crummy software is the amount of employment it generates. If Moore's Law is upheld for another twenty or thirty years, not only will there will be a vast amount of computation going on on Planet Earth, but also the maintenance of that computation will consume the efforts of almost every living person. We're talking about a planet of help desks.

I have argued elsewhere that this future would be a great thing, realizing the socialist dream of full employment by capitalist means. But let's consider the dark side.

Among the many processes that information systems make more efficient is the process of capitalism itself. A nearly friction-free economic environment allows fortunes to be accumulated in a few months instead of a few decades, but the individuals doing the accumulating are still living as long as they used to—longer, in fact. So those individuals who are good at getting rich have a chance to get richer before they die than their equally talented forebears. There are two dangers in this. The smaller, more immediate danger is that young people acclimatized to a deliriously receptive economic environment might be emotionally wounded by what the rest of us would consider brief returns to normalcy. The greater danger is that the gulf between the richest and the rest could become transcendently grave. That is, even if we agree that a rising tide raises all ships, if the rate of rising of the highest ships is greater than that of the lowest, they will become ever more separated. And indeed, concentrations of wealth and poverty increased during the Internet boom years in America. If Moore's Law or something like it is running the show, the scale of the separation could become astonishing. This is where my Terror resides, in considering the ultimate outcome of the increasing divide between the ultrarich and the merely better off.

With the technologies that exist today, the wealthy and the rest aren't all that different; both bleed when pricked, in the classic example. But with the technology of the next twenty or thirty years, they might become quite different indeed. Will the ultrarich and the rest even be recognizable as the same species by the middle of the new century? The possibilities that they will become essentially different species are so obvious and so terrifying that there is almost a banality in stating them. The rich could have their children made genetically more intelligent, beautiful, and joyous. Perhaps they could even be genetically disposed to have a superior capacity for empathy, but only to other people who meet some narrow range of criteria. Even stating these things seems beneath me, as if I were writing pulp science fiction, and yet the logic of the possibility is inescapable.

Let's explore just one possibility, for the sake of argument. One day the richest among us could turn nearly immortal, becoming virtual gods to the rest of us. (An apparent lack of aging in both cell cultures and in whole organisms has been demonstrated in the laboratory.) Let's not focus here on the fundamental questions of near immortality—whether it is moral or even desirable, or where one would find room if immortals insisted on continuing to have children. Let's instead focus on the question of whether immortality is likely to be expensive.

My guess is that immortality will be cheap if information technology gets much better and expensive if software remains as crummy as it is.

I suspect that the hardware/software dichotomy will reappear in biotechnology, and indeed in other twenty-first-century technologies. You can think of biotechnology as an attempt to make flesh into a computer, in the sense that biotechnology hopes to manage the processes of biology in ever greater detail, leading at some far horizon to perfect control. Likewise, nanotechnology hopes to do the same thing for materials science. If the body and the material world at large become more manipulable, more like a computer's memory, then the limiting factor will be the quality of the software that governs the manipulation.

Even though it's possible to program a computer to do virtually anything, we all know that's really not a sufficient description of computers. As I argued above, getting computers to perform specific tasks of significant complexity in a reliable but modifiable way, without crashes or security breaches, is essentially impossible. We can only approximate this goal, and only at great expense.

Likewise, one can hypothetically program DNA to make virtually any modification in a living thing, and yet designing a particular modification and vetting it thoroughly will likely remain immensely difficult. (And, as I argued above, that might be one reason why biological evolution has never found a way to be anything except very slow.) Similarly, one can hypothetically use nanotechnology to make matter do almost anything conceivable, but it will probably turn out to be much harder than we now imagine to get it to do any particular thing of complexity without distressing side effects. Scenarios predicting that biotechnology and nanotechnology will be able to quickly and cheaply create startling new things under the sun also must imagine that computers will become semiautonomous, superintelligent virtuoso engineers. But computers will do no such thing, if the last half-century of progress in software can serve as a predictor of the next half-century.

In other words, bad software will make biological hacks like near immortality expensive instead of cheap in the future. Even if everything else gets cheaper, the information technology side of the effort will get more expensive.

Cheap near immortality for everyone is a self-limiting proposition. There isn't enough room to accommodate such an adventure. Also, roughly speaking, if immortality were to become cheap, so would the horrific biological weapons of Bill Joy's scenario. On the other hand, expensive near immortality is something the world could absorb, at least for a good long while, because there would be fewer people involved. Maybe they could even keep the effort quiet.

So, here is the irony. The very features of computers which drive us crazy today and keep so many of us gainfully employed are the best insurance our species has for long-term survival as we explore the far reaches of technological possibility. Yet those same annoying qualities are what could make the twenty-first century a madhouse scripted by the fantasies and desperate aspirations of the superrich.

I share the belief of my cybernetic totalist colleagues that there will be huge and sudden changes in the near future brought about by technology. The difference is that I believe that whatever happens will be the responsibility of individual people who do specific things. I think that treating technology as if it were autonomous is the ultimate self-fulfilling prophecy. There is no difference between machine autonomy and the abdication of human responsibility.

Let's take the "nanobots take over" scenario. It seems to me that the most likely scenarios involve either:

a) Supernanobots everywhere that run old software—Linux, say. This might be interesting. Good video games will be available, anyway.
b) Supernanobots that evolve as fast as natural nanobots—so don't do much for millions of years.
c) Supernanobots that do new things soon but are dependent on humans. In these cases, humans will be in control, for better or for worse.

So, therefore, I'll worry about the future of human culture more than I'll worry about the gadgets. And what worries me about the Young Turk cultural temperament seen in cybernetic totalists is that they seem not to have been educated in the tradition of scientific skepticism. I understand why they are intoxicated. There *is* a compelling simple logic behind their thinking, and elegance in thought is infectious.

There is a real chance that evolutionary psychology, artificial intelligence, Moore's Law fetishizing, and the rest of the package will catch on in a big way—as big as Freud or Marx did in their time. Or bigger, since those ideas might end up essentially built into the software that runs our society and our lives. If that happens, the ideology of cybernetic totalist intellectuals will be amplified from a novelty into a force that could cause suffering for millions of people.

The greatest crime of Marxism wasn't simply that much of what it claimed was false but that it claimed to be the sole and utterly complete path to understanding life and reality. Cybernetic eschatology shares with some of history's worst ideologies a doctrine of historical predestination. There is nothing more gray, stultifying, or dreary than a life lived inside the confines of a theory. Let us hope that the cybernetic totalists learn humility before their day in the sun arrives.

PART III
EVOLVING UNIVERSES . . .

HOW FAST, HOW SMALL, AND HOW POWERFUL? MOORE'S LAW AND THE ULTIMATE LAPTOP

SETH LLOYD

Now we have created devices called computers, which can register and process huge amounts of information—a significant fraction of the amount of information that human beings themselves, as a species, can process. When I think of all the information being processed in that way . . . I see our species at a very interesting point in its history, which is the point at which our artifacts will soon be processing more information than we physically will be able to process.

SETH LLOYD is a professor of mechanical engineering at MIT and a principal investigator at MIT's Research Laboratory of Electronics. He works on problems having to do with information and complex systems, from the very small (How do atoms process information? How can you make them compute?) to the very large (How does society process information? And how can we understand society in terms of its ability to process information?). He is the author of *Programming the Universe*.

Computation is pervading the sciences. This encroachment seems to have begun about 400 years ago, if you look at the first paragraph of Hobbes's *Leviathan*. He says that just as we consider the human body to be like a machine—like a clock, where you have sinews and muscles to move energy about, a pulse beat like a pendulum, and a heart that pumps energy in the same way that a weight supplies energy to a clock's pendulum—then we can consider the State to be analogous to the body, since the State has a prince at its head, people who form its individual parts, legislative bodies that form its organs, and so forth. In that case, Hobbes asks, can we not consider the State itself to have an artificial life?

To my knowledge, that was the first use of the phrase "artificial life" in the way we use it today. If we have a physical system that's evolving in a physical way according to a set of rules, couldn't we consider it to be artificial and yet living?

Hobbes wasn't talking about information processing explicitly, but the examples he used were, in fact, examples of information processing. He used the example of the clock as something designed to process information: It gives you information about time. Most pieces of the clock he described are devices not only for transforming energy but for providing information. For example, the pendulum gives you regular, temporal information. When he discusses the State and imagines it having an artificial life, he first talks about "the head," the seat of the State's thought processes, and that analogy, in my mind, accomplishes two things. First, Hobbes is implicitly interested in information; second, he is constructing the fundamental metaphor of scientific and technological inquiry. When we think of a machine as possessing a kind of life in and of itself, and when we think of machines as doing the same kinds of things that we do, we are also thinking the corollary—that is, we are doing the same kinds of things that machines do. This metaphor, one of the most powerful of the Enlightenment, pervaded the popular culture of that time. Eventually, one could argue, it gave rise to Newton's dynamical picture of the world and the great inquiries into thermodynamics and heat that came 150 years later and became the central mechanical metaphor that informed all of science up to the twentieth century.

When did people first start talking about information in such terms that information processing, rather than clockwork, became the central metaphor for our times? People hadn't started thinking about the mechanical metaphor until they began building machines and had some good mechanical examples—like clocks, for instance. The seventeenth century was a fantastic century for clockmaking; in fact, the seventeenth and eighteenth centuries were fantastic centuries for building machines, period. Just as people began conceiving of the world using mechanical metaphors only when they had built machines, they began to conceive of the world in terms of information and information processing only when they began dealing with information and information processing. All the mathematical and theoretical materials for thinking of the world in terms of information were available at the end of the nineteenth century, because all the basic formulas had been created by James Clerk Maxwell, Ludwig Boltzmann, and Willard Gibbs for statistical mechanics. The formula for information was known back in the 1880s, but people didn't know that it dealt with information; instead, because they were familiar with things like heat and mechanical systems that processed heat, they called information, in its mechanical or thermodynamic manifestation, "entropy."

Then came the notion of constructing machines that actually processed information. Back in the early nineteenth century, Charles Babbage tried to construct one, but it was a spectacular and expensive failure and did not enter the popular mainstream. It wasn't until the 1930s that people like Claude Shannon and Norbert Weiner, and before them Harry Nyquist, started to think about information processing for purposes of communication, feedback, and control—a field that later became known as cybernetics. The notion in the late 1950s and early 1960s that cybernetics would solve all our problems—allowing us, for example, to figure out such matters as how our social systems work—was a colossal failure, not because that idea was necessarily wrong but because the techniques for doing so didn't exist at that point (and, if we're realistic, may in fact never exist). The successful applications of cybernetics are not even called cybernetics today, because they're so ingrained in our technology in fields like control theory and in the aerospace techniques that were used to put men on the moon. This brings us to the twenty-first century and the Internet, which in some sense is almost the evil twin of cybernetics. The word "cybernetics" comes from the Greek word *kybernotos*, which means a governor—helmsman, actually. The *kybernotos* was the pilot of a ship. Cybernetics, as initially conceived, was about governing, or controlling, or guiding. The great thing about the Internet, in my view, is that it's completely out of control.

I'm a physicist by training and I was taught to think of the world in terms of energy, momentum, pressure, entropy. You've got all this energy, things are happening, things are pushing on other things, things are bouncing around. But that's only half the story. The other half of the story, its complementary half, is the story about information, and here cybernetics was fundamentally on the right track. In one way you can think about what's going on in the world as energy, things moving around, things bouncing off one other—that's the way people have thought about the world for over 400 years, since Galileo and Newton. But what's missing from that picture is what that stuff is *doing*. That's a question about information. What's going on? Thinking about the world in terms of information is complementary to thinking about the world in terms of energy. To my mind, that's where the action is—thinking about the world as the confluence of information and energy and how they play off against each other. That's exactly what cybernetics was about. Norbert Weiner, the father of cybernetics, conceived of it in terms of information, things like feedback control. How much information, for example, do you need to make something happen? The first people studying these problems were scientists who happened to be physicists, and the first person who was clearly aware of the connection between information, entropy, and physical quantities such as energy was Maxwell, who in the 1850s and 1860s wrote down formulas that related what we would now call information to things like energy and entropy.

What chiefly separates humanity from most other living things is the way we deal with information. Somewhere along the line, perhaps around 100,000 years ago, we developed natural language, which is a universal method for processing information, and since then the history of human beings has been the development of ever-more sophisticated ways of registering, processing, transforming, and dealing with information. Now we have created devices called computers, which can register and process huge amounts of information—a significant fraction of the amount of information that human beings themselves, as a species, can process. When I think of all the information being processed in that way—all the information being communicated over the Internet; the total amount being processed by human beings and their artifacts—I see our species at a very interesting point in its history, which is the point at which our artifacts will soon be processing more information than we physically will be able to process. So I have to ask, How many bits am I processing per second in my head? I can estimate it. There are around 10 billion neurons, each one processing around 100 bits every 1/1000 of a second, so the total information processing power of the brain is around a million billion bits per second. If you believe that information processing is where the action is, it may mean that human beings are not going to be where the action is anymore. But because we created the devices that are doing this massive information processing, we, as a species, are uniquely poised to make our lives interesting and fun in completely unforeseen ways.

Every physical system, just by existing, can register information. And every physical system, just by evolving according to its own peculiar dynamics, can process that information. I'm interested in how the world registers information and how it processes it. But since I'm a scientist who deals with the physics of how things process information, I'm interested in that notion in a more specific way. I want to figure out not only how the world processes information but how *much* information it's processing. I've recently been working on methods to assign numerical values to how much information is being processed, just by ordinary physical dynamics. A few years ago, I got the idea of asking this question: Given the fundamental limits on how the world is put together—(1) the speed of light, which limits how rapidly information can get from one place to another; (2) Planck's constant, which tells you what the quantum scale is, how small things can actually get before they disappear altogether; and (3) the last fundamental constant of nature, the gravitational constant, which essentially tells you how large things can get before they collapse on themselves—how much information can possibly be processed? It turns out that the difficult part of this question was thinking it up in the first place. Once I'd managed to pose the question, it took me only six months to a year to figure out how to answer it, because the physics involved is pretty straightforward. It involves quantum mechanics, gravitation, and perhaps a bit of quantum gravity thrown in, but not enough to make things too difficult.

The other motivation for trying to answer this question was to analyze Moore's Law. Many of our society's prized objects are the products of this remarkable law of miniaturization. People have been getting *extremely* good at making the components of systems *extremely* small. This is what's behind the incredible increase in the power of computers, what's behind the amazing increase in information technology and communications (like the Internet), and what's behind pretty much every advance in technology you can think of, including fields like materials science. I like to consider it the most colossal land grab in the history of mankind.

From an engineering perspective, there are two ways to make something bigger. One is to make it physically bigger, and human beings spent a lot of time making things physically bigger—working out ways to deliver more power to systems, build bigger buildings, expand their territory at others' expense, and so on. But there's another way to make things bigger and that's to make things smaller. The real size of a system is not how big it actually is; the real size is the ratio between the biggest part of a system and the smallest part of a system—that is, the smallest part of a system that you can actually put to use. For instance, the reason computers are so much more powerful today than they were ten years ago is that every year and a half or so the basic components of computers (wires, logic chips, and so forth) have gone down in size by a factor of two. This is known as Moore's Law, which is just a historical fact about our technology.

Every time something's size goes down by a factor of two, you can cram twice as many of them into a box, and so every two years or so the power of computers doubles, and over the course of fifty years the power of computers has gone up by a factor of a million or more. The world has, in this sense, gotten a million times bigger, because we've been able to make the smallest usable parts of the world a million times smaller. So we are living in exciting times. But a reasonable question to ask is, Where will it all end? Since Moore proposed his law in the early 1960s, it has been written off numerous times. It was written off in the early 1970s because people thought that fabrication techniques for integrated circuits were going to break down and you wouldn't be able to get things smaller than a scale size of ten microns. Now

Moore's Law is being written off again, because people say that the insulating barriers between wires in computers are getting to be only a few atoms thick, and when you have an insulator only a few atoms thick, then electrons can tunnel through it and it's not a very good insulator. Well, perhaps that will stop Moore's Law—but nothing has stopped it so far.

At some point, does Moore's Law have to stop? This question involves the ultimate physical limits to computation: You can't send signals faster than the speed of light, you can't make things smaller than the laws of quantum mechanics permit, and if you make things too big then they just collapse into a giant black hole. As far as we know, it's impossible to fool Mother Nature.

I thought it would be interesting to see what the basic laws of physics say about how fast, how small, and how powerful computers can get. Actually these two questions— "Given the laws of physics, how powerful can computers be?" and "Where must Moore's Law eventually stop?"—turn out to be the same, because they're answered at the same place, which is where every available physical resource—every little subatomic particle, every ounce of energy, every photon in your system—is used to perform computation. The question is, How much computation is that? In order to investigate this, I thought a reasonable form of comparison would be to look at what I call the Ultimate Laptop. Let's ask how powerful this computer could be.

The idea here is that we can relate the laws of physics and the fundamental limits of computation to something we're familiar with—something of human scale that has a mass of about a kilogram, like a nice laptop computer, and has about a liter in volume, because kilograms and liters are pretty good to hold in your lap and a reasonable size to look at, to put in your briefcase, et cetera. After working on this problem for nearly a year, I was able to show that the laws of physics give absolute answers to how much information you can process with a kilogram of matter confined to a volume of one liter. The amount of information that can be processed, the number of bits you can register in the computer, and the number of operations per second you can perform on these bits are all related to basic physical quantities: to the aforementioned constants of nature—the speed of light, Planck's constant, and the gravitational constant. In particular, you can show without much trouble that the number of basic logical operations per second that you can perform using a certain amount of matter is proportional to the energy of this matter.

For those readers who are technically minded, it's not very difficult to whip out the famous formula $e = mc^2$ and show, using the work of Norman Margolus and Lev Levitin, that the total number of elementary logical operations you can perform per second using a kilogram of matter is the amount of energy, mc^2, times 2, divided by h-bar (Planck's constant), times pi. Well, you don't have to be Einstein to do the calculation. The mass is 1 kilogram, the speed of light is 3×10^8 meters per second, so mc^2 is about 10^{17} joules. Quite a lot of energy—I believe it's roughly the amount of energy used by all the world's nuclear power plants in the course of a week or so. A lot of energy, but let's suppose you could use it to do a computation. So you've got 10^{17} joules; and h-bar is 10^{-34} joules per second, roughly. So I have 10^{17} joules, I divide by 10^{-34} joules-seconds, and I have the number of operations: 10^{51} ops per second. So you can perform 10^{51} ops per second— and 10^{51} is about a million billion billion billion billion billion ops per second, a lot faster than conventional laptops.

And that's the answer. You can't do any better than that, as far as the laws of physics are concerned.

I wrote this up for *Nature* a couple of years ago, and ever since people keep calling me up to order one of these laptops. Unfortunately the fabrication plant to build them has not yet been constructed. You might well ask, "Why are our conventional laptops so slow by comparison, when we've been on this Moore's Law track for some fifty years now?" The answer is that conventional computers make the mistake (which could be regarded as a safety feature of the laptop) of locking up most of their energy in the form of matter, so that rather than using that energy to manipulate information and transform it, most of it goes into making the laptop sit around and be a laptop. If I were to take a week's energy output of all the world's nuclear power plants and liberate it all at once, I would have something that looked a lot like a thermonuclear explosion, because a thermonuclear explosion is essentially taking about a kilogram of matter and turning it into energy. So you can see right away that the ultimate laptop would have severe packaging and fabrication problems. It would not be easy to prevent this thing from taking not only you but the entire city of Boston out with it when you boot it up the first time.

Needless to say, I didn't figure out how we were going to put this ultimate laptop into a package, but that's part of the fun of doing calculations according to the ultimate laws of physics. I decided to calculate how many ops per second we could perform and to worry about the packaging afterward. Now that we've got 10^{51} ops per second, the next question is, What's the memory space of this laptop?

When I go out to buy a new laptop, I first ask how many ops per second it can perform. If it's something like 100 megahertz, it's pretty slow by current standards; if it's a gigahertz, that's pretty fast, though we're still very far away from 10^{51} ops per second. With a gigahertz, we're approaching 10^{10}, 10^{11}, 10^{12}, depending how ops per second are currently counted. Next, how many bits do I have? How big is the hard drive for this computer, or how big is its RAM? We can also use the laws of physics to calculate that figure, and computing memory capability is something that people could have done back in the early decades of this century.

We know how to count bits. We take the number of states, and the number of states is two raised to the power of the number of bits. Ten bits, 2^{10} states. You keep on going until you find that with about 300 bits—2^{300}—that's about 10^{100} states, which is essentially somewhat greater than the number of particles in the universe. If you had 300 bits, you could assign every particle in the universe a serial number, which is a powerful use of information. You could use a very small number of bits to label a huge number of objects.

How many bits does this ultimate laptop have?

I have a kilogram of matter confined to the volume of a liter. How many states—how many possible states for matter confined to the volume of a liter—can there possibly be? This happened to be a calculation that I know how to do, because I studied cosmology, and in cosmology there's this event called the Big Bang, which happened about 13 billion years ago. And during the Big Bang, matter was at extremely high densities and pressures. I learned from cosmology how to calculate the number of states for matter of very high densities and pressures. Of course, the density of our laptop is not that great; it's a kilogram of matter in a liter. However, if you want to ask what the number of states is for this matter in a liter, I've got to calculate every possible configuration, every possible elementary quantum state for this kilogram of matter in a liter of volume. It turns out that when you count most of these states, this matter looks like it's in the midst of a thermonuclear explosion, like a little piece of the Big Bang a few seconds after the universe was born, when the temperature was around a billion degrees. At a billion degrees, if you ask what most states for matter are, if it's completely liberated and able to do whatever it wants you'll find that it looks a lot like a plasma at a billion degrees Kelvin. Electrons and positrons are forming out of nothing and going back into photons again, and there are a lot of elementary particles whizzing about, and it's very hot. Lots of stuff is happening, but you can still count the number of possible states using the conventional methods that people use to count states in the early universe. You take the logarithm of the number of states, and you get a quantity that's normally thought of as being the entropy of the system. This will give you the number of bits; we find that there are roughly 10^{31} bits available. That means there's 2 to the 10^{31} possible states that this matter could be in. That's a lot of states, but we can count them. The interesting thing about that is that we've got 10^{31} bits, we're performing 10^{51} ops per second, so each bit can perform about 10^{20} ops per second. What does this quantity mean?

It turns out that this quantity—if you like, the number of ops per second per bit—is essentially the temperature of the plasma. I take this number and multiply it by Planck's constant, and what I get is the energy per bit, essentially. That's what temperature is; it tells you the energy per bit. It tells you how much energy is available for a bit to perform a logical operation. Since I know that if I have a certain amount of energy I can perform a certain number of operations per second, then the temperature tells me how many ops per bit per second can be performed by the Ultimate Laptop, a kilogram of matter in a liter volume: It's the number of ops per bit per second that could be performed by those elementary particles back at the beginning of time, in the Big Bang—the number of times a bit can flip, the number of times it can interact with its neighboring bits, the number of elementary logical operations. And it's a number, right? 10^{20}. Just as the total number of bits, 10^{31}, is a number—a physical parameter that characterizes a kilogram of matter and a liter of volume. Similarly, 10^{51} ops per second is the number of ops per second that characterizes a kilogram of matter, whether it's in a liter volume or not.

We've gone a long way down this road, so there's no point in stopping—at least in these theoretical exercises, where nobody gets hurt. So far, all we've used are the elementary constants of nature: the speed of light, which tells us how much energy we get from a particular mass, and the Planck scale, which tells us both how many operations per second you can get from a certain amount of energy and how to count the number of states available for a certain amount of energy. Thus we can calculate the number of ops per second that a certain amount of matter can perform and the amount of memory space that we have available for our ultimate computer.

Then we can also calculate all sorts of interesting issues, like what's the possible input/output rate for all these bits in a liter of volume. We can say, "OK, here's all these bits, they're sitting in a liter volume, let's move this liter volume over, by its own distance, at the speed of light." You're not going to be able to get the information in or out faster than that. And we find that we can get around 10^{40} or 10^{41} bits per second in and out of our Ultimate Laptop. That tells you how fast a modem you could possibly have for it: how many bits per second you could get in and out over the Ultimate Internet, whatever the Ultimate Internet would be. I guess the Ultimate Internet is just spacetime itself, in this picture.

I noted that you can't possibly do better than this; these are the laws of physics. But you might be able to do better in other ways. For example, let's think about the architecture of this computer. It's doing 10^{51} ops per second, with 10^{31} bits. Each bit can flip 10^{20} times per second. That's pretty fast. The next question is, How long does it take a bit on this side of the computer to send a signal to a bit on that side of the computer in the course of doing an operation?

As we've established, this computer has a liter volume, which is about ten centimeters on each side, so it takes about 10^{-10} seconds—one ten-billionth of a second—for light to go from one side to the other. The bits are flipping 100 billion billion times per second, so this bit is flipping 10 billion times in the time it takes a signal to go from one side of the computer to the other. But this is not a very serial computation; a lot of action is taking place over here in the time it takes to communicate with the other side of the computer. This is what's called parallel computation. You could say that—in the kinds of densities we're familiar with, like a kilogram per liter volume, which is the density of water—we can perform only a very parallel computation if we were to operate at the ultimate limits of computation; lots of computational action takes place over here during the time it takes a signal to go from here to there and back again.

How can we do better? How could we make the computation more serial?

Let's suppose we want our machine to do a more serial computation, so that in the time it takes to send a signal from one side of the computer to the other there are fewer ops being done. The obvious solution is to make the computer smaller, because if I make the computer smaller by a factor of two, it takes only half the time for light—that is, for a signal, for information—to go from one side of the computer to the other. If I make it smaller by a factor of 10 billion, it takes only a ten-billionth of the time for a signal to go from one side of the computer to the other. You also find that when you make the computer smaller, these pieces of the computer tend to speed up, because you tend to have more energy per bit available in each case. If you go through the calculation, you find out that as the computer gets smaller and smaller, as all the mass is compressed into a smaller and smaller volume, you can do a more serial computation.

When does this process stop? When can every bit in the computer talk with every other bit in the course of time it takes for a bit to flip? When can everybody get to talk with everybody else in the same amount of time that it takes them to talk with their neighbors?

As you make the computer smaller and smaller, it gets denser and denser; you have a kilogram of matter in an ever smaller volume. Eventually the matter assumes more and more interesting configurations, until it's actually going to take a very high pressure to keep this system down at this very small volume. The matter assumes stranger and stranger configurations and tends to get hotter and hotter and hotter, until at a certain point a bad thing happens. It's no longer possible for light to escape from it. It has become a black hole.

What happens to our computation at this point? This is probably very bad for computation, right? Or rather, it's going to be bad for input/output. Input is good, because stuff goes in, but output is bad, because it doesn't come out, since this is a black hole. Luckily, however, we're safe in this, because the laws of quantum mechanics we were using to calculate how much information a physical system can compute, how fast it can perform computations, and how much information it can register, still hold.

Stephen Hawking showed in the 1970s that black holes, if you treat them quantum-mechanically, can radiate out information. There's an interesting controversy as to whether that information has anything to do with the information that went in. Hawking and the Caltech theoretical physicist John Preskill have a well-known bet: Preskill says yes, the information that comes out of a black hole reflects the information that went in; Hawking says no, the information that comes out of a black hole when it radiates doesn't have anything to do with the information that went in, and the information that went in goes away. I don't know the answer to this.

But let's suppose for a moment that Hawking is wrong and Preskill is right. This kilogram black hole will be radiating at a whopping rate; it's radiating photons with wavelengths of 10^{-27} meters, which is not something you would actually wish to be close to. In fact, it would look a lot like a huge explosion. But let's suppose that the information being radiated by the black hole is in fact the information that went in to construct it in the first place, but that it is simply transformed in a particular way. What you then see is that the black hole can be thought of, in some sense, as performing a computation.

You take the information about the matter that's used to form the black hole, you program it (in the sense that you give it a particular configuration, you put this electron here, you put that electron there, you make that thing vibrate like this), and then you collapse this into a black hole. So 10^{-27} seconds later—in 100 billion billionth of a second—the thing goes *cablooey!* and you get all this information out again, but now the information has been transformed by some unknown dynamics. In fact, we would need to know something like string theory or quantum gravity to figure out how the information has been transformed. But you can imagine that this black hole could in fact function as a computer. We don't know how to make it compute, but indeed it's taking in information, it's transforming the information in a systematic form according to the laws of physics, and then *poop!*—it spits it out again. And suppose you could somehow read information coming out of the black hole. You would indeed have performed the ultimate computation that you could perform using a kilogram of matter, in this case confining it to a volume of 10^{-81} cubic meters.

Is there anything more to the story? After sending my paper on the Ultimate Laptop to *Nature*, I realized I had been insufficiently ambitious—that the obvious question to ask is not "What is the ultimate computational capacity of a kilogram of matter?" but "What is the ultimate computational capacity of the universe as a whole?" After all, the universe is processing information, is it not? Just by existing, all physical systems register information. Just by evolving their own natural physical dynamics, they transform that information, they process it. The real question is, How much information has the universe processed since the Big Bang?

A GOLDEN AGE
OF COSMOLOGY

ALAN GUTH

The classical theory was never really a theory of a bang; it was a theory about the aftermath of a bang. It started with all of the matter in the universe already in place, already undergoing rapid expansion, already incredibly hot. There was no explanation of how the universe got that way. Inflation is an attempt to answer the question of what made the universe bang, and now it looks as though it's almost certainly the right answer.

ALAN GUTH, the father of the inflationary theory of the universe, is the Victor F. Weisskopf Professor of Physics at MIT. His research interests are in the area of elementary particle theory and the application of particle theory to the early universe. In 2002 he was awarded the Dirac Medal of the International Centre for Theoretical Physics, along with Paul Steinhardt and Andrei Linde, for the development of the concept of inflation in cosmology. He is the author of *The Inflationary Universe*.

It's often said—and I believe this saying originated with the late astrophysicist David Schramm—that we are in a golden age of cosmology. That's really true. Cosmology is undergoing a transition from being a collection of speculations to becoming a genuine branch of hard science, in which theories can be developed and tested against precise observations. One of the most interesting areas is the prediction of the fluctuations, or nonuniformities, in the cosmic background radiation. We think of this radiation as the afterglow of the heat of the Big Bang; it's uniform in all directions to an accuracy of about 1 part in 100,000 after you subtract the term related to the motion of the earth through the background radiation.

I've been heavily involved in a theory called the inflationary universe, which seems to be our best explanation for this uniformity. The uniformity is hard to understand. You might think initially that it could be explained by the same physical principles that cause a hot slice of pizza to cool when you take it out of the oven: Things tend to come to a uniform temperature. But once the equations of cosmology were worked out so that one could calculate how fast the universe was expanding at any given time, physicists were able to calculate how much time there was for this uniformity to set in. They found that in order for the universe to have become uniform fast enough to account for the uniformity we see in the cosmic background radiation, information would have had to have been transferred at approximately 100 times the speed of light. But according to all our theories of physics, nothing can travel faster than light. So the classical version of the Big Bang theory had to simply assume that the universe was homogeneous—completely uniform—from the very beginning.

The inflationary-universe theory is an add-on to the standard Big Bang theory, and basically what it adds on is a description of what drove the universe into expansion in the first place. In the classical version of the Big Bang theory, that expansion was part of the initial assumptions; there's no explanation for it whatsoever. The classical theory was never really a theory of a bang; it was a theory about the aftermath of a bang. It started with all of the matter in the universe already in place, already undergoing rapid expansion, already incredibly hot. There was no explanation of how the universe got that way. Inflation is an attempt to answer the question of what made the universe bang, and now it looks as though it's almost certainly the right answer. It explains not only what caused the universe to expand, but also the origin of essentially all the matter in the universe at the same time. I say "essentially" because in a typical version of the theory, inflation needs about a gram's worth of matter to start. So inflation is not quite a theory of the ultimate beginning, but it is a theory of evolution that explains what we see around us starting from *almost* nothing.

Inflationary theory takes advantage of results from modern particle physics, which predicts that at very high energies there should exist peculiar kinds of substances that turn gravity on its head and produce repulsive gravitational forces. The inflationary explanation is the idea that the early universe contains at least a patch of this peculiar substance. All you need is a patch; it can actually be more than a billion times smaller than a proton. But once such a patch exists, its own gravitational repulsion causes it to grow, rapidly becoming large enough to encompass the entire observed universe.

From the time Einstein originally proposed it, general relativity has predicted the possibility of repulsive gravity; in the context of general relativity, you need a material with a negative pressure to create repulsive gravity. According to general relativity, it's not just matter densities or energy densities that create gravitational fields; it's also pressures. A positive pressure creates a normal, attractive gravitational field, of the kind we're accustomed to. A negative pressure would create a repulsive kind of gravity. It turns out, according to modern particle theories, that materials with a negative pressure are easy to construct out of fields that exist according to these theories. By putting together these two ideas—that particle physics gives us states with negative pressures and general relativity tells us that those states cause a gravitational repulsion—we reach the origin of the inflationary theory.

The inflationary theory gives a simple explanation for the uniformity of the observed universe, because in the inflationary model the universe starts out incredibly tiny. There would have been plenty of time for such a tiny region to reach a uniform temperature and uniform density, by the same mechanisms through which the air in a room reaches a uniform density throughout the room. In the early, tiny universe that the inflationary model posits, there would have been enough time for these mechanisms to cause an almost perfect uniformity. Then inflation takes over and magnifies this tiny region to become large enough to encompass the entire universe, maintaining this uniformity as the expansion takes place.

For a while, when the theory was first developed, we were worried that we would get too much uniformity. One of the amazing features of the universe is how uniform it is, but it's still by no means completely uniform. We have galaxies and stars and clusters and all kinds of complicated structures in the universe that need to be explained. If the universe started out completely uniform, it would have remained completely uniform, as there would be nothing to cause matter to collect here or there or any particular place.

Following earlier suggestions by Chibisov and Mukhanov, Stephen Hawking was one of the first people to explore what we now think is the answer to this riddle. He pointed out—although his first calculations were inaccurate—that quantum effects could come to our rescue. The real world is not fully described by classical physics, yet we were describing things completely classically, with deterministic equations. The real world, according to what we understand about physics, is described quantum mechanically, which means, deep down, that everything has to be described in terms of probabilities. The classical world we perceive, in which every object has a definite position and moves in a deterministic way, is really just the average of the possibilities that the full quantum theory predicts. If you apply that notion here, it is at least qualitatively clear from the beginning that it gets us in the direction we want to go—that is, the uniform density predicted by our classical equations would really be just the average of the quantum mechanical densities, which would have a range of values differing from one place to another. Quantum mechanical uncertainty would make the density of the early universe a little bit higher in some places and a little bit lower in other places. There would be fluctuations, so at the end of inflation we would expect to have ripples on top of an almost uniform density of matter.

It's possible to calculate these ripples. I should confess that we don't yet know enough about the particle physics to determine their amplitude, the intensity of the ripples, but what we *can* calculate is the way in which the intensity depends on the wavelength of the ripples. That is, there are ripples of all sizes, and you can measure the intensity of ripples of different sizes. And you can discuss what we call the spectrum—we use that word exactly the way it's used to describe sound waves. When we talk about the spectrum of a sound wave, we're talking about how the intensity varies with the different wavelengths that make up that sound wave. We do exactly the same thing in the early universe; we can talk about how the intensity of the ripples in the mass density of the early universe varied with their wavelengths.

Today we can see those ripples in the cosmic background radiation. The fact that we can see them at all is an absolute triumph of modern technology. When we were first making these predictions, back in 1982, astronomers had just barely begun to see the effect of the earth's motion through the background radiation, an effect of about 1 part in 1,000. The ripples I'm talking about are only 1 part in 100,000—just 1 percent of the intensity of the most subtle effect it had been possible to observe when we were first doing these calculations. I never believed we would actually see the ripples in the cosmic background; the idea that astronomers would get to be 100 times better at measuring these things just seemed too far-fetched. But to my astonishment and delight, in 1992 these ripples were detected by a satellite called COBE, the Cosmic Background Explorer. And now we have far better measurements than the ones made by COBE, which had an angular resolution of about 7 degrees, allowing us to see only the ripples of longest wavelength. Now we have measurements that go down to a fraction of a degree, and we're getting very precise measurements now of how the intensity varies with wavelength.

In the spring of 2000, there was a spectacular set of announcements from experiments called BOOMERANG (which stands for "Balloon Observations Of Millimetric Extragalactic Radiation ANd Geophysics") and MAXIMA (Millimeter Anisotropy eXperiment IMaging Array), both balloon-based experiments that provided very strong evidence that the universe is geometrically flat, which is just what inflation predicts. By "flat," we mean that the three-dimensional space of the universe is not curved, as it could have been according to general relativity. In the context of relativity, Euclidean geometry is not the norm—it's an oddity. In general relativity, curved space is the generic case. Once we assume that the universe is on average homogenous (the same in all places) and isotropic (the same in all directions), then the issue of flatness becomes directly related to the relationship between the universe's mass density and its expansion rate. A large mass density would cause space to curve into a closed universe, a universe in the shape of a ball; if the mass density dominates, the universe will be a closed space with a finite volume and no edge, in which a spaceship traveling in what it thinks is a straight line for a long enough

distance will end up back where it started from. In the alternative case, if the expansion dominates, the universe will be
geometrically open. Geometrically open spaces have the opposite geometric properties from closed spaces. They're infinite. In a closed space, two parallel lines will start to
converge; in an open space, two parallel lines will start to diverge. In either case, what you see is very different from Euclidean geometry. However, if the mass density is right at the
borderline of these two cases (known as the *critical density*),
then the geometry is Euclidean and the space is called flat.

The fact that the universe is at least approximately flat today requires that the early universe be extraordinarily flat.
The universe tends to evolve away from flatness, so even
given what we knew ten or twenty years ago, which is a good
deal less than what we know today, we could have extrapolated backward and discovered that, for example, at one second after the Big Bang the mass density of the universe must
have been at the critical density *to an accuracy of fifteen decimal
places*, in order to counterbalance the expansion rate and produce a flat universe. Conventional Big Bang theory provides
no mechanism to explain how the mass density became so
close to critical, but that's how things had to have been to explain why the universe looks the way it does today. Conventional Big Bang theory without inflation really works only if
you feed into it initial conditions that are highly fine-tuned
in order to produce a universe like the one we see. Inflationary theory gets around this so-called flatness problem, because inflation changes the way the geometry of the universe
evolves with time. Even though the universe always evolves
away from flatness at all other periods in its history, during
the inflationary period the universe is actually driven toward
flatness incredibly quickly. Inflation would have needed only

approximately 10^{-34} seconds or so to have driven the universe close enough to flatness to explain what we see today.

The inflationary mechanism that drives the universe toward flatness will in almost all cases overshoot, giving us not a universe that is close to flat today but a universe that is *almost exactly* flat today. Various people have tried to design versions of inflation that avoid this, but these versions seem contrived, requiring inflation to end at just the point where it's almost made the universe flat but not quite. The generic inflationary model drives the universe toward complete flatness, which means that one of its predictions is that today *the mass density of the universe should be at the critical value that makes the universe geometrically flat.* But until four or five years ago, no astronomers believed that. They told us that if you looked at just the visible matter, you would see only about 1 percent of what you needed to make the universe flat. They did offer more than that, in the form of "dark matter." Dark matter is matter whose existence is inferred from the gravitational effect it exerts on visible matter. Its effects are seen, for example, in the rotational curves of galaxies. When astronomers first measured how fast galaxies rotate, they found galaxies were spinning so fast that if the only matter present were the visible matter they would fly apart. The astronomers thus had to assume that there was a large amount of dark matter in a galaxy—about five or ten times the amount of its visible matter—holding the galaxy together. The same is true of the motion of galaxies within clusters of galaxies. That motion is much more random and chaotic than that of a single spiral galaxy, but you can nevertheless ask how much mass is needed to hold the clusters together, and the answer is that you still need significantly more matter than what you assumed was in the galaxies.

Adding all that up, astronomers came up with about a third of the critical density, and they were pretty well able to guarantee that there wasn't anymore stuff out there. That was bad for the inflationary model, but many of us still had faith in it and believed that sooner or later the astronomers would come up with something.

And they did. Starting in 1998, observations indicated the remarkable fact that the expansion of the universe appears to be accelerating, not slowing down. The theory of general relativity allows for that; what's needed is a material with a negative pressure. Most cosmologists are now convinced that our universe must be permeated by a material with negative pressure that is causing the acceleration we're now seeing. We don't know what this material is; we refer to it as "dark energy." Even though we don't know what it is, we can use general relativity itself to calculate how much mass has to be out there to cause the observed acceleration, and this number turns out to be almost exactly equal to two-thirds of the critical density—just what was missing from the previous calculations! On the assumption that this dark energy is real, we now have complete agreement between what the astronomers are telling us the mass density of the universe is and what inflation predicts.

At the time of the BOOMERANG and MAXIMA announcements, however, there was an important discrepancy that people worried about, and no one was sure how big a deal to make out of it. The spectrum they were measuring was a graph that had, in principle, several peaks. These peaks had to do with successive oscillations of the density waves in the early universe and a phenomenon called resonance that makes some wavelengths more intense than others. The measurements showed the first peak exactly where we expected it to be, with just the shape that was expected. But we couldn't see the second peak. In order to fit the data with theory, we had to assume that there were about ten times as many protons in the universe as we had thought, because those extra protons would lead to a friction effect that could make the second peak disappear. Of course, there is some uncertainty in any experiment; if an experiment is performed many times, the results will not be exactly the same each time. So we could comfort ourselves with the idea that the second peak was invisible purely because of bad luck; however, the probability that the peak could be *that* invisible if the universe contained the density of protons indicated by our other measurements was down in the 1 percent range. So here was a very serious discrepancy between what was observed and what was expected. All this has since changed dramatically for the better, with the next set of announcements of more precise measurements. Now the second peak is not only visible but is exactly as high as expected, and everything about the data now fits beautifully with the theoretical predictions. Too beautifully, really. I'm sure things will get worse before they continue to get better, given the difficulties in making these kinds of measurements. But we have a picture now that seems to confirm the inflationary theory of the early universe.

At the present time in our golden age of cosmology, the inflationary theory, which a few years ago was in significant conflict with observation, now works perfectly with our measurements of the mass density and the fluctuations. The evidence for a theory that's either the one I'm talking about or something very close to it is very, very strong. I should emphasize in closing that although I've been using the term in the singular, "inflation" is really a class of theories. If inflation is right, it's by no means the end of our study of the origin of the universe; it's really closer to the beginning. There are many different versions of inflation, and in fact the cyclic model that Paul Steinhardt describes in these pages could be considered one version—a rather novel version, since it puts the inflation at a completely different era of the history of the universe, but inflation is still doing many of the same things. There are many versions of inflation that are much closer to the kinds of theories we were developing in the 1980s and 1990s, so saying that inflation is "right" is by no means the end of the story. There's a lot of flexibility here, and a lot to be learned. And what needs to be learned will involve both the study of cosmology and the study of the underlying particle physics essential to these models.

THE CYCLIC UNIVERSE

PAUL STEINHARDT

[F]or the past year I've been involved in the development of an alternative theory that turns cosmic history topsy-turvy. In it, all the events that created the important features of our universe occur in a different order, by different physics, at different times, over different time scales—and yet this model seems capable of reproducing all of the successful predictions of the consensus picture with the same exquisite detail.

PAUL STEINHARDT is the Albert Einstein Professor in Science and a professor in both the Department of Physics and the Department of Astrophysical Sciences at Princeton University. He is one of the leading theorists responsible for inflationary theory, having been involved in constructing the first workable model of inflation and the theory of how inflation could produce seeds for galaxy formation. He was also among the first to show evidence for dark energy and cosmic acceleration, introducing the term "quintessence" to refer to dynamical forms of dark energy. In 2002 he was awarded the Dirac Medal of the International Centre for Theoretical Physics, along with Alan Guth and Andrei Linde, for the development of the concept of inflation in cosmology. He is co-author of *Endless Universe*.

If you were to ask most cosmologists to give a summary of where we stand right now in the field, they would tell you that we live in a very special period in human history, in which, thanks to a whole host of advances in technology, we can view the very distant and very early universe in ways we haven't been able to before. We can get a snapshot of what the universe looked like in its infancy, when the first atoms were forming. We can get a snapshot of what the universe looked like in its adolescence, when the first stars and galaxies were forming. And we are now getting a full-detail three-dimensional image of what the local universe looks like today. When you put this information together, you obtain a very tight series of constraints on any model of cosmic evolution. The data we've gathered in the last decade have eliminated all of the theories of cosmic evolution of the early 1990s save one—a model you might call today's consensus model. This model involves a combination of the Big Bang model as developed in the 1920s, 1930s, and 1940s; the inflationary theory, proposed by Alan Guth in the early 1980s; and a recent amendment I will discuss shortly. This consensus theory matches the observations we have of the universe today in exquisite detail. For this reason, many cosmologists conclude that we have finally determined the basic cosmic history of the universe.

But I have a rather different point of view, a view that has been stimulated by two events. The first is the recent amendment to which I referred. I want to argue that the recent amendment is not simply an amendment but a real shock to our whole notion of time and cosmic history. And second, for the past year I've been involved in the development of an alternative theory that turns cosmic history topsy-turvy. In it, the events that created the important features of our universe occur in a different order, by different physics, over a different time scale—and yet this model seems capable of reproducing all the successful predictions of the consensus picture with the same exquisite detail.

The key difference between this picture and the consensus picture comes down to the nature of time. The standard model, or consensus model, assumes that time has a beginning, which we normally refer to as the Big Bang. According to this model, for reasons we don't quite understand, the universe sprang from nothingness into somethingness, full of matter and energy, and has been expanding and cooling for the past 13.7 billion years. In the alternative model, the universe is endless. Time is endless in the sense that it goes on forever in the past and forever in the future. And in some sense, space is endless. Indeed, our three spatial dimensions remain infinite throughout the evolution of the universe.

More specifically, this model proposes a universe in which the evolution of the universe is cyclic. That is to say, the universe goes through periods of evolution from hot to cold, from dense to under-dense, from hot radiation to the structure we see today and eventually to an empty universe. Then a sequence of events occurs that causes the cycle to begin again. The empty universe is reinjected with energy, creating a new period of expansion and cooling. This process repeats periodically forever. What we're witnessing now is simply the latest cycle.

The notion of a cyclic universe is not new. People have considered this idea as far back as the beginning of recorded history. The ancient Hindus, for example, had a very elaborate and detailed cosmology based on a cyclic universe. They predicted the duration of each cycle to be 8.64 billion years—a prediction with three-digit accuracy. This is very impressive, especially since they had no quantum mechanics and no string theory! It disagrees with the number I'm going to suggest, which is trillions of years rather than billions.

The cyclic notion has also been a recurrent theme in western thought. Edgar Allan Poe and Friedrich Nietzsche, for example, each had cyclic models of the universe, and in the early days of relativistic cosmology Albert Einstein, Alexandr Friedman, Georges Lemaître, and Richard Tolman were all interested in the cyclic idea. I think it is clear why so many have found the cyclic idea to be appealing: If you have a universe with a beginning, you have the challenge of explaining why it began and the conditions under which it began. If you have a universe that is cyclic, it's eternal, so you don't have to explain the beginning.

During the attempts to try to bring cyclic ideas into modern cosmology in the 1920s and 1930s, various technical problems were discovered. The idea at that time was a cycle in which our three-dimensional universe goes through periods of expansion beginning with the Big Bang and then reversal to contraction and a Big Crunch. The universe bounces and expansion begins again. One problem is that every time the universe contracts to a Crunch, the density and temperature of the universe rises to an infinite value, and it is not clear that the usual laws of physics can be applied. Second, every cycle of expansion and contraction creates entropy through natural thermodynamic processes, which adds to the entropy from earlier cycles. So at the beginning of a new cycle, there is higher entropy density than the cycle before. It turns out that the duration of a cycle is sensitive to the entropy density. If the entropy increases, the duration of the cycle increases as well. So, going forward in time, each cycle becomes longer than the one before. The problem is that extrapolating back in time the cycles become shorter and shorter, until after a finite time they shrink to zero duration. The problem of avoiding a beginning has not been solved; it has simply been pushed back a finite number of cycles. If we're going to reintroduce the idea of a truly cyclic universe, these two problems must be overcome. The cyclic model I will describe uses new ideas to do just that.

To appreciate why an alternative model is worth pursuing, it's important to get a more detailed impression of the consensus picture. Certainly some aspects of the consensus model are appealing, but recent observations have forced us to amend the consensus model and make it more complicated. So let me begin with an overview of it.

The consensus theory begins with the Big Bang: The universe has a beginning. It's an assumption that people have made over the last fifty years, but it's not something we can prove at present from any fundamental laws of physics. Furthermore, you have to assume that the universe began with an energy density less than the critical value. Otherwise the universe would have stopped expanding and would have recollapsed before the next stage of evolution, the inflationary epoch. Moreover, to reach this inflationary stage there must be some sort of energy to drive the inflation. Typically this is assumed to be due to an "inflaton" field. You have to assume that in those patches of the universe that began at less than the critical density, a significant fraction of the energy was stored in inflaton energy, which could eventually overtake the universe and start the period of accelerated expansion. All these are reasonable assumptions, but they are assumptions nonetheless.

Assuming these conditions are met, the inflaton energy overtakes the matter and radiation after a few instants. The inflationary epoch commences and the expansion of the universe accelerates at a furious pace. The inflation does a number of miraculous things: It makes the universe homogeneous, it makes the universe flat, and it leaves behind certain inhomogeneities, which are supposed to be the seeds for the formation of galaxies. Now the universe is prepared to enter the next stage of evolution with the right conditions. According to the inflationary model, the inflaton energy decays into a hot gas of matter and radiation. After a second or so, the first light nuclei form. After a few tens of thousands of years, the slowly moving matter dominates the universe. It is during this period that the first atoms form, the universe becomes transparent, and the structure in the universe begins to form—the first stars and galaxies. Up to this point the story is relatively simple.

But there is the recent discovery that we have entered a new stage in the evolution of the universe. Something strange has happened to cause the expansion of the universe to speed up again. During the 13.7 billion years when matter and radiation dominated the universe and structure was forming, the expansion of the universe was slowing down, because the matter and radiation within it gravitationally attractive and resist the expansion. Until very recently, it was presumed that matter would continue to be the dominant form of energy in the universe and that this deceleration would continue forever.

But we've discovered instead, in recent observations, that the expansion of the universe is speeding up. This means that most of the energy of the universe is neither matter nor radiation. Rather, another form of energy has overtaken the matter and radiation. For lack of a better term, this new energy form is called "dark energy." Dark energy, unlike the matter and radiation we're familiar with, is gravitationally repulsive. That's why it causes the expansion to speed up rather than slow down. In Newton's theory of gravity, all mass is gravitationally attractive, but Einstein's general theory of relativity allows the possibility of forms of energy that are gravitationally self-repulsive.

I don't think either the physics or cosmology communities, or even the general public, have fully absorbed the implications of this discovery. This is a revolution in the grand historic sense—in the Copernican sense. Copernicus (from whom we derive the word "revolution") changed our notion of space and our position in the universe. By showing that the earth revolves around the sun, he triggered a chain of ideas that led us to the notion that we live in no particular place in the universe; there's nothing special about where we are. Now we've discovered something very strange about the nature of time: that we may live in no special place, but we *do* live at a special time. It is a time of recent transition from deceleration to acceleration; from one in which matter and radiation dominate the universe to one in which they are rapidly becoming insignificant components; from one in which structure is forming on ever-larger scales to one in which, because of this accelerated expansion, structure formation stops. We are in the midst of the transition between these two stages of evolution. And just as Copernicus's proposal that the earth is no longer the center of the universe led to a chain of ideas that changed our outlook on the structure of the solar system and eventually the structure of the universe, this new discovery of cosmic acceleration may well lead to a change in our view of cosmic history.

With these thoughts about the consensus model in mind, let me turn to the cyclic proposal. Since it's cyclic, I'm allowed to begin the discussion of the cycle at any point I choose. To make the discussion parallel, I'll begin at a point analogous to the Big Bang; I'll call it the Bang. This is a point in the cycle where the universe reaches its highest temperature and density. In this scenario, though, unlike the *Big* Bang model, the temperature and density don't diverge. There is a maximal, finite temperature. It's a very high temperature—around 10^{20} degrees Kelvin, hot enough to evaporate atoms and nuclei into their fundamental constituents—but it's not infinite. In fact, it's well below the so-called Planck energy scale, where quantum gravity effects dominate. The theory begins with a Bang and then proceeds directly to a phase dominated by radiation. In this scenario you do not have the inflation you have in the standard scenario. You still have to explain why the universe is flat, you still have to explain why the universe is homogeneous, and you still have to explain where the fluctuations came from that led to the formation of galaxies, but that's not going to be explained by an early stage of inflation. It's going to be explained by yet a different stage in the cyclic universe, which I'll get to.

In this new model, the universe proceeds directly to a radiation-dominated phase and forms the usual nuclear abundances; then it goes directly to a matter-dominated phase, in which the atoms and galaxies and larger-scale structure form; and then it proceeds to a phase dominated by dark energy. In the consensus model, the dark energy comes as a surprise, since it is something you have to add into the theory to make it consistent with what we observe. In the cyclic model, the dark energy moves to center stage as the key player that drives the universe into the cyclic evolution. The first thing the dark energy does when it dominates the universe is what we observe today: It causes the expansion of the universe to begin to accelerate. Why is that important? Although this acceleration rate is 100 orders of magnitude smaller than the acceleration one has in inflation, if you give the universe enough time this slow accelerated expansion actually accomplishes the same feat that inflation does. Over time, it thins out the distribution of matter and radiation in the universe, making the universe more and more homogeneous and isotropic—in fact, making it perfectly so, driving it into what is essentially a vacuum state.

There are 10^{80} or 10^{90} particles inside the universal horizon volume (13.7 billion light-years in radius), but if you look around the universe in a trillion years you will find, on average, the particles have been spread out so far that there is less than one particle within that same volume. Seth Lloyd wants us to view the universe as a computer, in which the bits—the particles—available for computation are those within the horizon. In an accelerating universe, Seth's ultimate computer is actually losing bits.

At the same time that the universe is made homogeneous and isotropic, it is also being made flat. If the universe had any warp or curvature to it, although it's a slow process, the acceleration caused by the dark energy makes the space extremely flat. If accelerated expansion continued forever, of course, that would be the end of the story. But in this scenario, just like inflation, the dark energy survives only for a finite period. Then it triggers a series of events that eventually lead to a transformation of energy from gravity potential energy into new energy and radiation, which will then start a new period of expansion of the universe. This rapid production of matter and radiation and the associated reversal from contraction to rapid expansion constitute the next bang. From a local observer's point of view, it looks like the universe goes through exact cycles; that is to say, it looks like the universe empties out on each round and a new matter and radiation is created, leading to a new period of expansion. In this sense, it's a cyclic universe.

If you were a global observer and could see the entire infinite universe, you'd discover that our three dimensions are forever infinite in this story. What happens is that each time matter and radiation are created, they get thinned out by a large but finite factor. They're out there somewhere but getting thinned out. Locally it looks like the universe is cyclic, but globally the universe has a steady evolution, in which the total entropy increases by a constant factor from one cycle to the next. Extrapolating backward in time, the universe is contracting and the entropy is decreasing by a constant factor each cycle. However, if the universe is infinite and the total entropy is infinite, then decreasing by a finite factor still leaves an infinite volume and infinite entropy. In principle, the process could continue indefinitely.

Exactly how this works in detail can be described in various ways. I choose to present a very nice geometrical picture motivated by superstring theory. We use only a few basic elements from superstring theory, so you don't really have to know anything about superstring theory to understand what I'm going to talk about, except to understand that some of the strange things I'm going to introduce are already part of superstring theory, waiting to be put to some use.

One of the ideas in superstring theory is that there are extra dimensions; it's an essential element, necessary to make the theory mathematically consistent. In one particular formulation of that theory, the universe has a total of eleven dimensions. Six of them are curled up into a little ball so tiny that, for my purposes, I'm just going to pretend they're not there. However, there are three spatial dimensions, one time dimension, and one additional dimension I do want to consider. In this picture, our three spatial dimensions lie along a hypersurface, or membrane. This membrane is a boundary of the extra spatial dimension. There is another boundary or membrane forming the other boundary of the extra dimension. In between, in the so-called *bulk-volume*, there is the extra dimension, which, unlike our usual three dimensions, stretches only a finite interval. It's as if our three-dimensional world were one face of a sandwich with another three-dimensional world on the other face. These faces are referred to as *orbifolds* or *branes* (the latter from the word "membrane"). The branes have physical properties. They have energy and momentum, and when you excite them you can produce quarks and electrons. All of us are composed of the quarks and leptons on our branes. And since quarks and leptons can move only along branes, we are restricted to moving along and seeing only the three dimensions of our brane. We cannot see directly the bulk or any matter on the other brane.

In the cyclic universe, at regular intervals of trillions of years, these two branes smash together. This creates all kinds of excitations—particles and radiation. The collision thereby heats up the branes and then they bounce apart again.

The branes are attracted to each other by a force that acts just like a spring, causing them to come together at regular intervals. During each cycle, the universe goes through two kinds of motion. When the universe has matter and radiation in it, the main motion is that the branes are stretching—or, equivalently, our three dimensions are expanding. At the same time, the branes remain more or less a fixed distance apart. This period spans the 13.7 billion years since the last bang. The stretching is what we normally interpret as the expansion of the universe. All during this period, at a microscopic distance away, there is another brane sitting and expanding, but since we can't touch, feel, or see across the bulk, we can't sense it directly. If there is a clump of matter over there, we can feel its gravitational effect, but we can't see any light or anything else it emits, because anything it emits is going to move along that brane. We see only things that move along our own brane.

Next, when the radiation and matter are thinned out, the energy associated with the force between these branes takes over the universe. From our vantage point on one of the branes, this acts just like the dark energy we observe today. It causes the branes to accelerate in their stretching until all the matter and radiation produced since the last collision is spread out and the branes become essentially smooth, flat, empty surfaces. If you like, you can think of them as being wrinkled and full of matter after 13.7 billion years, and then stretching by a fantastic amount over the next trillion years. The stretching causes the mass and energy on our brane to thin out and the wrinkles to be smoothed. After trillions of years, the branes are, for all intents and purposes, smooth, flat, parallel, and empty.

Then the force between these two branes slowly brings them together. As it brings them together, the force grows stronger and the branes speed toward each other. When they collide, there's a walloping impact—enough to create a high density of matter and radiation with a very high, albeit finite temperature. The two branes go flying apart, more or less back to where they are now, and then the new matter and radiation (through the action of gravity) cause the branes to begin a new period of stretching.

In this picture, it's clear that the universe is going through periods of expansion and a funny kind of contraction. Where the two branes come together, it's not a contraction of our dimensions but a contraction of the extra dimension. Before that contraction, all matter and radiation have been spread out, but, unlike the old cyclic models of the 1920s and 1930s, they don't come back together again during the contraction, because our three-dimensional world—that is, our brane—remains stretched out. Only the extra dimension contracts. This process repeats itself, cycle after cycle.

If you compare the cyclic model to the consensus picture, two of the functions of inflation—namely, flattening and homogenizing the universe—are accomplished by a period of accelerated expansion like that we've now just begun. Of course, this flattening and homogenizing took place well before the present galaxies of our universe were formed, so it is the analogous expansion that occurred one cycle ago, before the most recent bang, that made our universe homogeneous and flat. Once this occurred, it remained nearly homogeneous and flat as the universe contracted and reexpanded full of matter and radiation.

The third function of inflation—producing fluctuations in the density—occurs as these two branes come together and the extra dimension contracts. As they approach, quantum fluctuations cause the branes to begin to wrinkle. And because they are wrinkled, they do not collide everywhere at the same time. Rather, some regions collide a bit earlier than others. This means that some regions reheat to a finite temperature and begin to cool a little bit before other regions. When the branes come apart again, the temperature of the universe is not perfectly homogeneous but has tiny spatial variations in temperature and density left over from the quantum wrinkles.

Remarkably, although the physical processes that create fluctuations in the cyclic model are completely different physically and have a completely different time scale from that of the inflationary model—taking billions of years instead of 10^{-30} seconds—it turns out that the spectrum of fluctuations in the distribution of energy and temperature generated in the two cosmological models is essentially identical. Hence, the cyclic model is also in exquisite agreement with the measurements of universal temperature and mass distribution we have today.*

The physical process that generates the fluctuations in the two models results in a subtle but important distinction, which can be checked by future experiments. In inflation, fluctuations in spacetime itself, called gravitational waves,

---

*On February 11, 2003, the Wilkinson Microwave Anisotropy Probe (WMAP) satellite team announced its landmark results, presenting a highly precise snapshot of the distribution of temperature and energy in the very early universe. The team emphasized the comparison with the consensus Big Bang/inflationary model, reporting that the simplest inflationary models are now ruled out. Less emphasized is the fact that the simplest cyclic models are consistent with their results. It is too early to reach strong conclusions from the current results, but they hint that we are at the threshold of the critical observations that may enable us to distinguish between the two scenarios.

are created in addition to the fluctuations in energy and temperature. That's a feature we hope to look for in experiments in the coming decades as a verification of the consensus model. In the cyclic model, you don't get those gravitational waves. The essential difference is that inflationary fluctuations are created in a hyperrapid violent process strong enough to create gravitational waves, whereas cyclic fluctuations are created in an ultraslow, gentle process too weak to produce gravitational waves. That's an example in which the two models give a dramatically different observational prediction. The gravitational wave signal is too difficult to observe at the present time, but experiments in the next decade may be sensitive enough.

What's fascinating at the moment is that we have two paradigms available to us. On the one hand, they are poles apart in terms of what they tell us about the nature of time, about our cosmic history, about the order in which events occur, and about the time scale on which they occur. On the other hand, they are remarkably similar in terms of what they predict about the universe today. Ultimately what will decide between the two is a combination of observation (for example, the search for cosmic gravitational waves) and theory, because a key aspect of the cyclic scenario entails assumptions about what happens at the collision between branes— assumptions that might well be checked or refuted in superstring theory. In the meantime, for the next few years, we can all have great fun speculating about the implications of each of these ideas—which we prefer and how we can best distinguish between them.

THEORIES OF
THE BRANE

LISA RANDALL

Additional spatial dimensions may seem like a wild and crazy idea at first, but there are powerful reasons to believe that there really are extra dimensions of space. One reason resides in string theory, in which it is postulated that the particles are not themselves fundamental but are oscillation modes of a fundamental string.

LISA RANDALL is a professor of physics at Harvard University, where she also earned her Ph.D. (1987). Between 1998 and 2000, she had a joint appointment at Princeton and MIT as a full professor, and she moved to Harvard as a full professor in 2001. Her research in theoretical high-energy physics is primarily related to exploring the physics underlying the standard model of particle physics. This has involved studies of supersymmetry and, most recently, extra dimensions of space. She is the author of *Warped Passages*.

———————

Particle physics has contributed to our understanding of many phenomena, ranging from the inner workings of the proton to the evolution of the observed universe. Nonetheless, fundamental questions remain unresolved, motivating speculations beyond what is already known. These mysteries include the perplexing masses of elementary particles; the nature of the dark matter and dark energy that constitute the bulk of the universe; and what predictions string theory, the best candidate for a theory incorporating both quantum mechanics and general relativity, makes about our observed world. Such questions (along with basic curiosity) have prompted my excursions into theories that might underlie currently established knowledge. Some of my most recent work has been on the physics of extra dimensions of space and has proved rewarding beyond expectation.

Particle physics addresses questions about the forces we understand—the electromagnetic force, the weak forces associated with nuclear decay, and the strong force that binds quarks together into protons and neutrons—but we still have to understand how gravity fits into the picture. String theory is the leading contender, but we don't yet know how string theory reproduces all the particles and physical laws we actually see. How do we go from this pristine, beautiful theory existing in ten dimensions to the world surrounding us, which has only four—three spatial dimensions plus time? What has become of string theory's superfluous particles and dimensions?

Sometimes a fruitful approach to the big, seemingly intractable problems is to ask questions whose possible answers will be subject to experimental test. These questions generally address physical laws and processes we've already seen. Any new insights will almost certainly have implications for even more fundamental questions. For example, we still don't know what gives rise to the masses of the fundamental particles—the quarks, leptons (the electron, for example), and electroweak gauge bosons—or why these masses are so much less than the mass associated with quantum gravity. The discrepancy is not small: The two mass scales are separated by sixteen orders of magnitude! Only theories that explain this huge ratio are likely candidates for theories underlying the standard model. We don't yet know what that theory is, but much of current particle physics research, including that involving extra dimensions of space, attempts to discover it. Such speculations will soon be explored at the Large Hadron Collider in Geneva, which will operate at the

TeV energies relevant to particle physics. The results of experiments to be performed there should select among the various proposals for the underlying physical description in concrete and immediate ways. If the underlying theory turns out to be either supersymmetry or one of the extra-dimension theories I will go on to describe, it will have deep and lasting implications for our conception of the universe.

Right now, I'm investigating the physics of the TeV scale. Particle physicists measure energy in units of electron volts. TeV means a trillion electron volts. This is a very high energy and challenges the limits of current technology, but it is low from the perspective of quantum gravity, whose consequences are likely to show up only at energies sixteen orders of magnitude higher. This energy scale is interesting because we know that the as-yet-undiscovered part of the theory associated with giving elementary particles their masses should be found there.

Two of the potential explanations for the huge disparity in energy scales are supersymmetry and the physics of extra dimensions. Supersymmetry, until very recently, was thought to be the only way to explain physics at the TeV scale. It is a symmetry that relates the properties of bosons to those of their partner fermions (bosons and fermions being two types of particles distinguished by quantum mechanics). Bosons have integral spin and fermions have half-integral spin, where spin is an internal quantum number. Without supersymmetry, one would expect these two particle types to be unrelated. But given supersymmetry, properties like mass and the interaction strength between a particle and its supersymmetric partner are closely aligned. It would imply for an electron, for example, the existence of a corresponding superparticle—called a selectron, in this case—with the same mass and charge. There was and still is a big hope that we will find signatures of supersymmetry in the next generation of colliders. The discovery of supersymmetry would be a stunning achievement. It would be the first extension of symmetries associated with space and time since Einstein constructed his theory of general relativity in the early twentieth century. And if supersymmetry is right, it is likely to solve other mysteries, such as the existence of dark matter. String theories that have the potential to encompass the standard model seem to require supersymmetry, so the search for supersymmetry is also important to string theorists. Both for these theoretical reasons and for its potential experimental testability, supersymmetry is a very exciting theory.

However, like many theories, supersymmetry looks fine in the abstract but leaves many questions unresolved when you get down to the concrete details of how it connects to the world we actually see. At some energy, supersymmetry must break down, because we haven't yet seen any "superpartners." This means that the two particle partners—for example, the electron and the selectron—cannot have exactly the same mass; if they did, we would see both. The unseen partner must have a bigger mass if it has so far eluded detection. We want to know how this could happen in a way consistent with all known properties of elementary particles. The problem for most theories incorporating supersymmetry-breaking is that all sorts of other interactions and decays are predicted which experiment has already ruled out. The most obvious candidates for breaking supersymmetry permit the various kinds of quarks to mix together, and particles would have a poorly defined identity. The absence of this mixing and the retention of the various quark identities are a stringent constraint on the content of the physical theory associated with supersymmetry-breaking, and are one important reason that people were not completely satisfied with super-symmetry as an explanation of the TeV scale. To find a consistent theory of supersymmetry requires introducing physics that gives masses to the supersymmetric partners of all the particles we know to exist, without introducing interactions we don't want. So it's reasonable to look around for other theories that might explain why particle masses are associated with the TeV energy scale and not one that is sixteen orders of magnitude higher.

There was a lot of excitement when it was first suggested that extra dimensions provide alternative ways to address the origin of the TeV energy scale. Additional spatial dimensions may seem like a wild and crazy idea at first, but there are powerful reasons to believe that there really are extra dimensions of space. One reason resides in string theory, in which it is postulated that the particles are not themselves fundamental but are oscillation modes of a fundamental string. The consistent incorporation of quantum gravity is the major victory of string theory. But string theory also requires nine spatial dimensions, which, in our observable universe, is obviously six too many. The question of what happened to the six unseen dimensions is an important issue in string theory. But if you're coming at it from the point of view of the relatively low-energy questions, you can also ask whether extra dimensions could have interesting implications in our observable particle physics or in the particle physics that should be observable in the near future. Can extra dimensions help answer some of the unsolved problems of three-dimensional particle physics?

People entertained the idea of extra dimensions before string theory came along, although such speculations were soon forgotten or ignored. It's natural to ask what would happen if there were different dimensions of space; after all, the fact that we see only three spatial dimensions doesn't necessarily mean that only three exist, and Einstein's general relativity doesn't treat a three-dimensional universe pre-ferentially. There could be many unseen ingredients to the universe. However, it was first believed that if additional dimensions existed they would have to be very small in order to have escaped our notice. The standard supposition in string theory was that the extra dimensions were curled up into incredibly tiny scales—10^{-33} centimeters, the so-called Planck length and the scale associated with quantum effects becoming relevant. In that sense, this scale is the obvious candidate: If there are extra dimensions, which are obviously important to gravitational structure, they'd be characterized by this particular distance scale. But if so, there would be very few implications for our world. Such dimensions would have no impact whatsoever on anything we see or experience.

From an experimental point of view, though, you can ask whether extra dimensions really must be this ridiculously small. How large could they be and still have escaped our notice? Without any new assumptions, it turns out that extra dimensions could be about seventeen orders of magnitude larger than 10^{-33} cm. To understand this limit requires more fully understanding the implications of extra dimensions for particle physics.

If there are extra dimensions, the messengers that potentially herald their existence are particles known as Kaluza-Klein modes. These KK particles have the same charges as the particles we know, but they have momentum in the extra dimensions. They would thus appear to us as heavy particles with a characteristic mass spectrum determined by the extra dimensions' size and shape. Each particle we know of would have these KK partners, and we would expect to find them if the extra dimensions were large. The fact that we have not yet seen KK particles in the energy regimes we have explored experimentally puts a bound on the extra dimensions' size. As I mentioned, the TeV energy scale of 10^{-16} cm has been explored experimentally. Since we haven't yet seen KK modes and 10^{-16} cm would yield KK particles of about a TeV in mass, that means all sizes up to 10^{-16} are permissible for the possible extra dimensions. That's significantly larger than 10^{-33} cm, but it's still too small to be significant.

This is how things stood in the world of extra dimensions until very recently. It was thought that extra dimensions might be present but that they would be extremely small. But our expectations changed dramatically after 1995, when Joe Polchinski, of the University of California at Santa Barbara, and other theorists recognized the importance of additional objects in string theory called "branes". Branes are essentially membranes—lower-dimensional objects in a higher-dimensional space. (To picture this, think of a shower curtain, virtually a two-dimensional object in a three-dimensional space.) Branes are special, particularly in the context of string theory, because there's a natural mechanism to confine particles to the brane; thus, not everything need travel in the extra dimensions even if those dimensions exist. Particles confined to the brane would have momentum and motion only along the brane, like water spots on the surface of your shower curtain.

Branes allow for an entirely new set of possibilities in the physics of extra dimensions, because particles confined to the brane would look more or less as they would in a three-plus-one-dimension world; they never venture beyond it. Protons, electrons, quarks, all sorts of fundamental particles could be stuck on the brane. In that case, you may wonder why we should care about extra dimensions at all, since despite their existence the particles that make up our world do not traverse them. However, although all known standard-model particles stick to the brane, this is not true of gravity. The mechanisms for confining particles and forces mediated by the photon or electrogauge proton to the brane do not apply to gravity. Gravity, according to the theory of general relativity, must necessarily exist in the full geometry of space. Furthermore, a consistent gravitational theory requires that the graviton, the particle that mediates gravity, has to couple to any source of energy, whether that source is confined to the brane or not. Therefore, the graviton would also have to be out there in the region encompassing the full geometry of higher dimensions—a region known as the "bulk"—because there might be sources of energy there. Finally, there is a string-theory explanation of why the graviton is not stuck to any brane: The graviton is associated with the closed string, and only open strings can be anchored to a brane.

A scenario in which particles are confined to a brane and only gravity is sensitive to the additional dimensions permits extra dimensions that are considerably larger than previously thought. The reason is that gravity is not nearly as well tested as other forces, and if it is only gravity that experiences extra dimensions, the constraints are much more permissive. We haven't studied gravity as well as we've studied most other particles, because it's an extremely weak force and therefore more difficult to precisely test. Physicists have showed that even dimensions almost as big as a millimeter would be permitted, if it were only gravity out in the higher-dimensional bulk. This size is huge compared with the scales we've been talking about. It is a macroscopic, visible size! But because photons (which we see with) are stuck to the brane, too, the dimensions would not be visible to us, at least in the conventional ways.

Once branes are included in the picture, you can start talking about crazily large extra dimensions. If the extra dimensions are very large, that might explain why gravity is so weak. (Gravity might not seem weak to you, but it's the entire earth that's pulling you down; the result of coupling an individual graviton to an individual particle is quite small. From the point of view of particle physics, which looks at the interactions of individual particles, gravity is an extremely weak force.) This weakness of gravity is a reformulation of the so-called hierarchy problem—that is, why the huge Planck mass suppressing gravitational interactions is sixteen orders of magnitude bigger than the mass associated with particles we see. But if gravity is spread out over large extra dimensions, its force would indeed be diluted. The gravitational field would spread out in the extra dimensions and consequently be very weak on the brane—an idea recently proposed by theorists Nima Arkani-Hamed, Savas Dimopoulos, and Gia Dvali. The problem with this scenario is the difficulty of explaining why the dimensions should be so large. The problem of the large ratio of masses is transmuted into the problem of the large size of curled-up dimensions.

Raman Sundrum, currently at Johns Hopkins University, and I recognized that a more natural explanation for the weakness of gravity could be the direct result of the gravitational attraction associated with the brane itself. In addition to trapping particles, branes carry energy. We showed that from the perspective of general relativity this means that the brane curves the space around it, changing gravity in its vicinity. When the energy in space is correlated with the energy on the brane so that a large, flat three-dimensional brane sits in the higher-dimensional space, the graviton (the particle communicating the gravitational force) is highly attracted to the brane. Rather than spreading uniformly in an extra dimension, gravity stays localized, very close to the brane.

The high concentration of the graviton near the brane—let's call the brane where gravity is localized the "Planck brane"—leads to a natural solution to the hierarchy problem in a universe with two branes. For the particular geometry that solves Einstein's equations, when you go out some distance in an extra dimension, you see an exponentially suppressed gravitational force. This is remarkable because it means that a huge separation of mass scales—sixteen orders of magnitude—can result from a relatively modest separation of branes. If we are living on the second brane (not the Planck brane), we would find that gravity was very weak. Such a moderate distance between branes is not difficult to achieve and is many orders of magnitude smaller than that necessary for the large-extra-dimensions scenario just discussed. A localized graviton plus a second brane separated from the brane on which the standard model of particle physics is housed provide a natural solution to the hierarchy problem—the problem of why gravity is so incredibly weak. The strength of gravity depends on location, and away from the Planck brane it is exponentially suppressed.

This theory has exciting experimental implications, since it applies to a particle physics scale—namely, the TeV scale. In this theory's highly curved geometry, Kaluza-Klein particles—those particles with momentum in the extra dimensions—would have mass of about a TeV; thus, there is a real possibility of producing them in colliders in the near future. They would be created like any other particle and they would decay in much the same way. Experiments could then look at their decay products and reconstruct the mass and spin that is their distinguishing property. The graviton is the only particle we know about that has spin 2. The many Kaluza-Klein particles associated with the graviton would also have spin 2 and could therefore be readily identified.

Observation of these particles would be strong evidence of the existence of additional dimensions and would suggest that this theory is correct.

As exciting as this explanation of the existence of very different mass scales is, Raman and I discovered something perhaps even more surprising. Conventionally, it was thought that extra dimensions must be curled up or bounded between two branes, or else we would observe higher-dimensional gravity. The aforementioned second brane appeared to serve two purposes: It explained the hierarchy problem because of the small probability for the graviton to be there, and it was also responsible for bounding the extra dimension so that at long distances (bigger than the dimension's size) only three dimensions are seen.

The concentration of the graviton near the Planck brane can, however, have an entirely different implication. If we forget the hierarchy problem for the moment, the second brane is unnecessary! That is, even if there is an infinite extra dimension and we live on the Planck brane in this infinite dimension, we wouldn't know about it. In this "warped geometry," as the space with exponentially decreasing graviton amplitude is known, we would see things as if this dimension did not exist and the world were only three-dimensional.

Because the graviton has such a small probability of being located away from the Planck brane, anything going on far away from the Planck brane should be irrelevant to physics on or near it. The physics far away is in fact so entirely irrelevant that the extra dimension can be infinite, with absolutely no problem from a three-dimensional vantage point. Because the graviton makes only infrequent excursions into the bulk, a second brane or a curled-up dimension isn't necessary to get a theory that describes our three-dimensional world, as had previously been thought. We might live on the Planck brane and address the hierarchy problem in some other manner—or we might live on a second brane out in the bulk, but this brane would not be the boundary of the now-infinite space. It doesn't matter that the graviton occasionally leaks away from the Planck brane; it's so highly localized there that the Planck brane essentially mimics a world of three dimensions, as though an extra dimension didn't exist at all. A four-spatial-dimensions world, say, would look almost identical to one with three spatial dimensions. Thus, all the evidence we have for three spatial dimensions could equally well be evidence for a theory in which there are four spatial dimensions of infinite extent.

It's an exciting but frustrating game. We used to think the easiest thing to rule out would be large extra dimensions, because large extra dimensions would be associated with low energies, which are more readily accessible. Now, however, because of the curvature of space, there is a theory permitting an infinite fourth dimension of space in a configuration that so closely mimics three dimensions that the two worlds are virtually indistinguishable.

If there are differences, they will be subtle. It might turn out that black holes in the two worlds would behave differently. Energy can leak off the brane, so when a black hole decays it might spit out particles into the extra dimension and thus decay much more quickly. Physicists are now doing some interesting work on what black holes would look like if this extra-dimensional theory with the highly concentrated graviton on the brane is true; however, initial inquiries suggest that black holes, like everything else, would look too similar to distinguish the four- and three-dimensional theories. With extra dimensions, there are an enormous number of possibilities for the overall structure of space. There can be

different numbers of dimensions and there might be arbitrary numbers of branes contained within. Branes don't even
all have to be three-plus-one-dimensional; maybe there are
other dimensions of branes in addition to those that look like
ours and are parallel to ours. This presents an interesting
question about the global structure of space, since how space
evolves with time would be different in the context of the
presence of many branes. It's possible that there are all sorts
of forces and particles we don't know about that are concentrated on branes and can affect cosmology.

In the above example, physics everywhere—on the brane
and in the bulk—*looks* three-dimensional. Even away from
the Planck brane, physics appears to be three-dimensional,
albeit with weaker gravitational coupling. Working with Andreas Karch (now at the University of Washington), I discovered an even more amazing possibility: Not only can there
be an infinite extra dimension, but physics in different locations can reflect different dimensionality. Gravity is localized
near us in such a way that it's only the region near us that
looks three-dimensional; regions far away reflect a higher-
dimensional space. It may be that we see three spatial dimensions not because there really are only three spatial
dimensions but because we're stuck to this brane and gravity
is concentrated near it, while the surrounding space is oblivious to our lower-dimensional island. There are also possibilities that matter can move in and out of this isolated
four-dimensional region, seeming to appear and disappear as
it enters and leaves our domain. These are very hard phenomena to detect in practice, but theoretically there are all
sorts of interesting questions about how such a construct all
fits together.

Whether these theories are right will not necessarily be
answered experimentally but could be argued for theoreti-

cally, if one or more of them ties into a more fundamental theory. We've used the basic elements found in string theory—namely, the existence of branes and extra dimensions—but we would really like to know if there is a true brane construction. Could you take the very specific branes given by string theory and produce a universe with a brane that localizes gravity? Whether you can actually derive this from string theory or some more fundamental theory is important. The fact that we haven't done it yet isn't evidence that it's not true, and Andreas and I have made good headway into realizing our scenario in string theory. But it can be very, very hard to solve these complicated geometrical setups. In general, the problems that get solved, although they seem very complicated, are in many ways simple problems. There is much more work to be done; exciting discoveries await, and they will have implications for other fields.

In cosmology, for instance. Alan Guth's mechanism whereby exponential expansion smooths out the universe works very well, but another possibility has been suggested: a cyclic universe, Paul Steinhardt's idea, wherein a smaller amount of exponential expansion happens many times. Such a theory prompts you to ask questions. First of all, is it really consistent with what we see? The jury's out on that. Does it really have a new mechanism in it? In some sense, the cyclic idea still uses inflation to smooth out the universe. Sometimes it's almost too easy to come up with theories. What grounds your theories? What ties them down? What restricts you from just doing anything? Is there really a new idea there? Do we really have a new mechanism at work? Does it connect to some other, more fundamental theoretical idea? Does it help make that work? Recently I have been exploring the implications of extra dimensions for cosmology. It seems that inflation with extra dimensions works even better than without! What's so nice about this theory is that one can reliably calculate the effect of the extra dimension; no *ad hoc* assumptions are required. Furthermore, the theory has definite implications for cosmology experiments. All along, I've been emphasizing what we actually see. It's my hope that time and experiments will distinguish among the possibilities.

LOOP QUANTUM GRAVITY

LEE SMOLIN

It's only since the middle 1980s that real progress began to be made on unifying relativity and quantum theory. The turning point was the invention of not one but two approaches: loop quantum gravity and string theory. Since then, we have been making steady progress on both of these approaches. In each case, we are able to do calculations that predict surprising new phenomena. Still, we are not done. Neither is yet in final form; there are still things to understand. But the really important news is that there is now a real chance of doing experiments that will test the new predictions of these theories.

This is important, because we're in the uncomfortable situation of having two well-developed candidates for the quantum theory of gravity. We need to reduce these to one theory. We can do this either by finding that one is wrong and the other right, or by finding that the two theories can themselves be unified.

LEE SMOLIN, a theoretical physicist, is concerned with quantum gravity, "the name we give to the theory that unifies all the physics now under construction." More specifically, he is a co-inventor of an approach called loop quantum gravity. In 2001, he became a founding member and research physicist of the Perimeter Institute for Theoretical Physics, in Waterloo, Ontario. Smolin is the author of *The Life of the Cosmos*, *Three Roads to Quantum Gravity*, and *The Trouble with Physics*.

Several years ago, I had the chance to move to Imperial College in London with the possibility of starting a research group. After I had been there awhile, someone came to see me and said, "I represent some people who want to start an institute for theoretical physics. They want it to do fundamental work in things like quantum gravity, string theory, cosmology, and quantum mechanics and they have at least $100 million. What would you do? What fields would be included? How would you structure it? Who would be good to hire? Would you have a director? Would you pick somebody honored and wise and give him all the power to structure it, or would you just hire a bunch of young people out of graduate school and give it to them on a high-tech entrepreneurial model and let them run with it?" We talked about this, and he also talked to many people in these fields— Fotini Markopoulou-Kalamara, Carlo Rovelli, Chris Isham, Roger Penrose, and many others.

Very important in these discussions was the structure. My view was that it's wrong to give one person all the power, because this is science, and science functions best when people are independent and there's a community. The proposed institute—the Perimeter Institute for Theoretical Physics, in Waterloo, just outside of Toronto—was particularly meant to be an incubator for innovative ideas about fundamental questions, and new ideas tend either to come from young people or from people who keep themselves young by constantly moving into new areas. We scientists constantly criticize each other, and we work best in an open atmosphere where anybody can criticize anybody, honestly and directly. You also want a supportive atmosphere, where people are generous and sympathetic about difficulties and failures. We talked about all these things, and over time the prospect began to look more attractive than staying in London.

The inventor of the idea, and the chief donor to the Perimeter Institute, is Michael Lazaridis, who is co-chief executive of Research in Motion, the company that makes Blackberries. He and the board he created made it clear that what they wanted, structurally, was something like the Institute for Advanced Study, in Princeton. They set the mandate, they set the framework, but they are not involved in day-to-day issues of scientific direction and hiring. Mike is absolutely essential, but he's never come to us and said, "I think you have to hire this person" or "I think that that's not a good direction to go in." One thing they did very early was to create a committee of prominent scientists as advisors, to oversee what we do. They're there to see that we don't wander off in strange directions scientifically—to keep us honest.

We're now in a funky old building in Waterloo which used to be a restaurant; my office is next to the old bar. There's a wonderful atmosphere; people love it. Construction has started on a new building designed by two fantastic young architects from Montreal, Gilles Saucier and Andre Perrotte. At the beginning of the process, we traveled with them to Cambridge and London, where people have recently built buildings for physicists and mathematicians, and talked about what works, what doesn't work, and why. I believe that our building is going to be a better place to do theoretical physics than anything that exists now. We already are said by some to be the hot place in two fields—quantum gravity and quantum information theory. We opened in September 2001, which was a strange time to begin any endeavor, starting with three scientists on long-term appointments: Robert Myers, Fotini Markopoulou, and myself—a string theorist and two people in quantum gravity. Very much present in our minds from the beginning was the idea that we were not going to favor one particular approach. We have good people in both camps, and we are creating an atmosphere where people in different camps will talk to each other. A lot of good science has happened so far. We hired two very good people in quantum theory: Lucien Hardy, from Oxford, who has done exciting work in foundations of quantum theory and quantum information theory; and Daniel Gottesman, a young star of quantum information theory. In 2002, we had ten postdocs, several visitors, lots of people coming and going. In June, the Canadian prime minister and the minister of industry visited and pledged more than $25 million to our support. The deputy provincial minister of Ontario also came and pledged at least $11 million. It was heartening to see that the leaders of at least one country understand that the support of pure science is essential for a modern democracy.

Science is a kind of open laboratory for a democracy. It's a way to experiment with the ideals of our democratic societies. For example, in science you must accept the fact that you live in a community that makes the ultimate judgment as to the worth of your work. But at the same time, everybody's judgment is his or her own. The ethics of the community require that you argue for what you believe and that you try as hard as you can to get results to test your hunches, but you have to be honest in reporting the results, whatever they are. You have the freedom and independence to do whatever you want, as long as in the end you accept the judgment of the community. Good science comes from the collision of contradictory ideas, from conflict, from people trying to do better than their teachers did, and I think here we have a model for what a democratic society is about. There's a great strength in our democratic way of life, and science is at the root of it.

Now I want to talk about the problem of quantum gravity and the two best-developed approaches that have been proposed to solve it, which are called loop quantum gravity and string theory. This is a case in which different people have taken different approaches to solving a fundamental scientific problem, and there are interesting lessons to be learned from how these theories have developed since the early 1980s—lessons about space and time and also about how science works.

Quantum gravity is the name we give to the theory that unifies all of physics. The roots of it are in Einstein's general theory of relativity and in quantum theory. Einstein's general theory of relativity is a theory of space, time, and gravity, while quantum theory describes everything else that exists in the universe, including elementary particles, nuclei, atoms, and chemistry. These two theories were invented in the early twentieth century, and their ascension marked the overthrow of the previous theory, which was Newtonian mechanics.

They are the primary legacies of twentieth-century physics. The problem of unifying them is the main open problem in physics left for us to solve in this century.

Nature is a unity. This pen is made of atoms *and* it falls in the earth's gravitational field. Hence there must be one framework, one law of nature of which these two theories are different aspects. It would be absurd if there were two irreconcilable laws of physics, one for one domain of the world and another for another domain. Even in 1915 Einstein was aware of the issue, and in his very first paper about gravitational waves, he mentions the paradox of how to fit relativity together with the quantum.

It's only since the middle 1980s that real progress began to be made on unifying relativity and quantum theory. The turning point was the invention of not one but two approaches: loop quantum gravity and string theory. Since then, we have been making steady progress on both of these approaches. In each case, we are able to do calculations that predict surprising new phenomena. Still, we are not done. Neither is yet in final form; there are still things to understand. But the really important news is that there is now a real chance of doing experiments that will test the new predictions of these theories.

This is important, because we're in the uncomfortable situation of having two well-developed candidates for the quantum theory of gravity. We need to reduce these to one theory. We can do this either by finding that one is wrong and the other right, or by finding that the two theories can themselves be unified. (Of course, the result of testing the theories could be that both of them are eliminated, but this would be progress, too.)

Until a few years ago, the situation was very different. We didn't know how to test the theories we were working so hard to construct. Indeed, for a whole scientific generation— that is, since the middle 1970s—fundamental physics has been in a crisis, because it has not been possible to subject our theoretical speculations to experimental test. This was because the new phenomena that our theories of quantum gravity predict occur at scales of energy many orders of magnitude greater than what can be created in the laboratory— even in the huge particle accelerators. The scale where quantum physics and gravity come together is called the Planck scale, and it is some fifteen orders of magnitude higher in energy than the largest accelerators now under construction.

In quantum theory, distance is inverse to energy, because you need particles of very high energy to probe very short distances. The inverse of the Planck energy is the Planck length. It is where the classical picture of space as smooth and continuous is predicted by our theories to break down, and it is some twenty powers of ten smaller than an atomic nucleus. Because the Planck scale is so remote from experiment, people began to put great trust in mathematics and theory. There were even some string theorists who said silly things like "From Galileo to 1984 was the period of modern physics, where we checked our theories experimentally. Since then, we work in the age of postmodern physics, in which mathematical consistency suffices to demonstrate the correctness of our theories and experiment is neither possible nor necessary." I'm not exaggerating; people really said things like this.

The idea that you could do experiments to test the quantum theory of gravity was mentioned from time to time by a few people through the 1990s, but to our shame we ignored them. One person who proposed the idea forcefully is a young man in Rome called Giovanni Amelino-Camelia. He just ignored everybody who said, "You'll never probe scales that small. You'll never test these theories." He told himself that there must be a way, and he examined many different possible experiments, looking for ways that effects of quantum gravity could appear. And he found them. Now we know more than half a dozen different experiments we can do to test different hypotheses about physics at the Planck scale. Indeed, in the last year, several proposals about Planck scale physics have been ruled out by experiment.

The key thing that Amelino-Camelia and others realized is that we can use the universe itself as an experimental device to probe the Planck scale. There are three ways the universe gives us experimental probes of the Planck scale. First, there are accelerators in distant galaxies that produce particles with energies much higher than we can produce in even the largest human-made accelerators. Some of these ultra-high-energy cosmic rays have been observed hitting our atmosphere with energies more than 10 million times those we have ever produced. These provide us with a set of ready-made experiments, because on their way to us they have traveled great distances through the radiation and matter that fill the universe. Indeed, there are already surprises in the data which, if they hold up, can be interpreted as due to effects of quantum gravity.

Second, we detect light and particles that have traveled billions of light-years on their way across the universe to us. During the billions of years they travel, very small effects due to quantum gravity can be amplified to the point that we can detect them.

Finally, the postulated inflation by which the universe expanded very rapidly at early times serves as a kind of microscope, blowing up Planck scale features to astronomical scales, where we can see them in small fluctuations in the cosmic microwave radiation.

So what are the theories we will be testing with these effects? One is loop quantum gravity.

Loop quantum gravity started in the early 1980s with some discoveries about classical general relativity by Amitaba Sen, then a postdoc at the University of Maryland. These were made into a beautiful reformulation of Einstein's theory by Abhay Ashtekar, then at Syracuse University and now director of the Center for Gravitational Physics at Penn State—a reformulation that brought the mathematical and conceptual language we use to describe space and time closer to the language used in particle physics and quantum physics. My colleague Ted Jacobson of the University of Maryland and I then found in 1986 that we could use this new formalism of Ashtekar's to get real results about quantum spacetime. Since the 1950s, the key equation of quantum gravity has been one called the Wheeler-DeWitt equation. Bryce DeWitt and John Wheeler wrote it down, but in all the time since then, no one had been able to solve it. We found we could solve it exactly, and in fact we found an infinite number of exact solutions. They revealed a microscopic structure to the geometry of space and told us that space, at the Planck scale, looks like a network with discrete edges joined into graphs. The next year, I was joined by Carlo Rovelli (now of the Centre de Physique Théorique in Marseille), and we were able to make a full-fledged quantum theory of gravity out of these solutions. This became loop quantum gravity. We were quickly joined by many others, and now it is a rather large field of research.

Loop quantum gravity differs from other approaches to quantum gravity, such as string theory, in that apart from using Ashtekar's formalism we made no modifications to the principles of relativity and quantum theory. These principles are well tested by experiment, and our theory is based on their consistent unification, nothing more. Our approach joins relativity in the world as we see it, with three spatial dimensions and matter more or less as we see it, with quantum mechanics more or less in the form presented to us by Paul Dirac, Werner Heisenberg, and their friends. While most people had given up and were seeking to modify the principles of either relativity or quantum theory, we surprised ourselves (and many other people) by succeeding in putting them together without modifying their principles.

This has led to a detailed theory that gives us a new picture of the nature of space and time as they appear when probed at the Planck scale. The most surprising aspect of this picture is that on that scale, space is not continuous but made up of discrete elements. There is a smallest unit of space: Its minimum volume is given roughly by the cube of the Planck length (which is 10^{-33} cm). A surface dividing one region of space from another has an area that comes in discrete units, the smallest of which is roughly the Planck length squared. Thus, if you take a volume of space and measure it to very fine precision, you discover that the volume can't be just anything. It has to fall into some discrete series of numbers, just like the energy of an electron in an atom. And just as in the case of the energy levels of atoms, we can calculate the discrete areas and volumes from the theory.

When we first worked out the predictions for these smallest units of area and volume, we had no idea that they would be observable in real experiments in our lifetime. However, a number of people—beginning with Rodolfo Gambini, of the University of the Republic in Montevideo, and Jorge Pullin, then at Penn State—showed that there are indeed observable consequences. At about the same time,

Amelino-Camelia and others were pointing out that if there were such effects, they would be detectable in experiments involving cosmic rays and gamma-ray bursts. These effects are caused by light scattering off the discrete structure of the quantum geometry, analogous to diffraction and refraction from light scattering off the molecules of the air or liquid it passes through. The quantum gravity effect is tiny—many orders of magnitude smaller than that due to matter. However, we observe light from gamma-ray bursts—huge explosions, possibly caused by mergers of binary neutron stars or black holes—that has traveled across the universe for some 10 billion light-years. Over such long distances, the small effects amplify to the point where they can be observed. Because elementary particles travel as waves in quantum theory, the same thing happens to such particles—protons and neutrinos, for example. It is possible that these effects may be responsible for the surprises I mentioned in the observations of very-high-energy cosmic rays.

Now, here is the really interesting part: Some of the effects predicted by the theory appear to be in conflict with one of the principles of Einstein's *special* theory of relativity, the theory that says that the speed of light is a universal constant. It's the same for all photons, and it is independent of the motion of the sender or observer.

How is this possible, if that theory is itself based on the principles of relativity? The principle of the constancy of the speed of light is part of special relativity, but we quantized Einstein's general theory of relativity. Because Einstein's special theory is only a kind of approximation to his general theory, we can implement the principles of the latter but find modifications to the former. And this is what seems to be happening!

So Gambini, Pullin, and others calculated how light travels in a quantum geometry and found that the theory predicts that the speed of light has a small dependence on energy. Photons of higher energy travel slightly slower than low-energy photons. The effect is very small, but it amplifies over time. Two photons produced by a gamma-ray burst 10 billion years ago, one redder and one bluer, should arrive on Earth at slightly different times. The time delay predicted by the theory is large enough to be detectable by a new gamma-ray observatory called GLAST (Gamma-ray Large Area Space Telescope), which is scheduled for launch into orbit in 2006. We very much look forward to the announcement of the results, as they will be testing a prediction of a quantum theory of gravity.

A very exciting question we are now wrestling with is, How drastically shall we be forced to modify Einstein's special theory of relativity if the predicted effect is observed? The most severe possibility is that the principle of relativity simply fails. The principle of relativity basically means that velocity is relative and there is no absolute meaning to being at rest. To contradict this would mean that after all there is a preferred notion of rest in the universe. This, in turn, would mean that velocity and speed are absolute quantities. It would reverse 400 years of physics and take us back before Galileo enunciated the principle that velocity is relative. While the principle may have been approximately true, we have been confronting the frightening possibility that the principle fails when quantum gravity effects are taken into account.

Recently, people have understood that this possibility appears to be ruled out by experiments that have already been done: that is, if the principle of relativity fails when quantum gravity effects are taken into account, effects would already have been seen in certain very delicate measurements involving atomic clocks and in certain astrophysical processes involving supernova remnants. These effects are not seen, so this drastic possibility seems less likely. So a hypothesis about the structure of space and time on scales twenty orders of magnitude smaller than an atomic nucleus has been ruled out by experiment!

But there is another possibility. This is that the principle of relativity is preserved, but Einstein's special theory of relativity requires modification so as to allow photons to have a speed that depends on energy. The most shocking thing I have learned in the last year is that this is a real possibility. A photon can have an energy-dependent speed without violating the principle of relativity! This was understood a few years ago by Amelino-Camelia. I got involved in this issue through work I did with João Magueijo, a very talented young cosmologist at Imperial College. During the two years I spent working there, João kept coming to me and bugging me with this problem. His reason for asking was that he had realized that if the speed of light could change according to conditions—for example, when the universe was very hot and dense—you might get an alternative cosmological theory. He and Andreas Albrecht (and before them John Moffat) had found that if the speed of light was higher in the early universe, you get an alternative to inflationary cosmology that explains everything inflation does, without some of the baggage.

These ideas all seemed crazy to me, and for a long time I didn't get it. I was sure it was wrong! But João kept bugging me and slowly I realized that they had a point. We have since written several papers together showing how Einstein's postulates may be modified to give a new version of special relativity in which the speed of light can depend on energy.

Meanwhile, in the last few years there have been some important new results concerning loop quantum gravity.

One is that the entropy of a black hole can be computed, and it comes out exactly right. Jacob Bekenstein found in his Ph.D. thesis in 1971 that every black hole must have an entropy proportional to the area of its horizon, the surface beyond which light cannot escape. Stephen Hawking then refined this by showing that the constant of proportionality must be, in units in which area is measured by the Planck length squared, exactly one quarter. A challenge for all quantum theories of gravity since then has been to reproduce this result. Moreover, entropy is supposed to correspond to a measure of information: It counts how many bits of information may be missing in a particular observation. So if a black hole has entropy, one has to answer the question, What is the information that the entropy of a black hole counts?

Loop quantum gravity answers these questions by giving a detailed description of the microscopic structure of the horizon of a black hole. This is based on the atomic description of spatial geometry, which implies that the area of a black hole horizon is quantized—just as space is, it is made up of discrete units. It turns out that a horizon can have, for each quantized unit of area, a finite number of states. Counting them, we get exactly Bekenstein's result, with the one quarter.

This is a very recent result. When we first did this kind of calculation, in the mid-1990s, we got the entropy right up to an overall constant. A few months ago, in a brilliant paper, Olaf Dreyer, a postdoc at the Perimeter Institute, found a very simple and original argument that fixes that constant, using a completely classical property of black holes. He uses an old argument of Neils Bohr called the correspondence principle, which tells us how to tie together classical and quantum descriptions of the same system. Once the constant is fixed, it gives the right entropy for all black holes.

Another big development of loop quantum gravity is that

we now know how to describe not only space but space-time—including causality, light cones, and so on—in loop quantum gravity. Spacetime also turns out to be discrete, described by a structure called a spin foam. Recently there have been important results showing that dynamical calculations in spin-foam models come out finite. Together these two results strongly suggest that loop quantum gravity is giving us sensible answers to questions about the nature of space and time on the shortest scales.

Let me now say something about string theory, which is the other approach to quantum gravity that has been well studied.

String theory is a very beautiful subject. It attempts to unify gravity with the other forces by postulating that all particles and forces arise from the vibrations of extended objects. These include one-dimensional objects (hence the name "strings"), but there are also higher-dimensional extended objects that go by the name of "branes" (for generalizations of membranes). String theory comes from the observation that all the quanta that carry the known forces, and all the known particles, can be found among the vibrations of these extended objects.

String theory is not a complete quantum theory of gravity, for reasons I'll come to in a minute, but it does work to a certain extent. It gives, to a certain order of approximation, sensible predictions for some quantum gravity effects. These include the scattering of gravitons (quanta of gravity analogous to photons) with other particles. For certain very limited kinds of black holes (actually, not real black holes but systems with properties similar to certain special black holes), it gives predictions that agree with the results of Bekenstein and Hawking. And it does succeed in unifying gravity with the other forces.

However, there is some fine print. For string theory to work, we need to hypothesize that there are six or seven unobservable dimensions of space. We must also hypothesize that there are new kinds of symmetries called supersymmetries, which have not so far been observed. These symmetries tie together particles usually considered constituents of matter (like quarks and electrons) with the quanta of forces (like photons and gluons).

Supersymmetry is a beautiful idea—and, indeed, it stands independent of string theory as an intriguing conjecture about the elementary particles. Unfortunately, it is not observed. Were it observed directly, then for every particle there would be a supersymmetric partner, which is a partner with the same mass and the same charges and interactions but a spin differing by one-half. This is certainly not observed! If supersymmetry is true, then it is realized in nature only indirectly; we say, in physicist talk, that the symmetry is broken. Another way to say this is that the forces have a symmetry, but the state of the world does not obey it. (For example, looking around your living room, you see that the fact that space has three-dimensional symmetry is broken by the effects of the gravitational field, which points down.)

There is some indirect evidence that some people take as an indication that supersymmetry is present and will be seen in future experiments in accelerators. But so far no direct evidence for supersymmetry has been found. Nor has there been any experimental evidence for the extra dimensions that string theory requires.

The interesting—and unfortunate—upshot of this is that in the absence of experimental check, different communities of people have focused on different questions and invented different imaginary worlds. Those who work on loop quantum gravity still live in the world we see, where space has three dimensions and there is no need for more symmetries than are observed. Many string theorists live—at least, imaginatively—in a universe that has ten or eleven dimensions. A standard joke is that a string theorist hearing a talk about loop quantum gravity says, "That's a very beautiful theory, but it has two big faults: Space only has three dimensions and there is no supersymmetry!" To which the speaker replies, "You mean, just like the real world?" Actually this is not a joke—I've heard it. (And, by the way, if the world does have higher dimensions and supersymmetry, that could be incorporated into loop quantum gravity.)

The extent to which people can invent imaginary worlds when science gets decoupled from experiment is quite extraordinary. They follow a certain aesthetic of mathematical elegance out there as far as it takes them. If you buy all that—the extra dimensions and symmetries and so forth—string theory does succeed to a certain limited approximation in unifying gravity and quantum theory. However, even if it's right, string theory can be only an approximation to the real theory. One reason is that there are an enormous number of string theories. And so far, while many of them have been studied, no single string theory has been discovered that agrees with all the observations of our universe. There are three features of the world that no string theory can so far reproduce: the absence of supersymmetry at low energies, the presence of a cosmological constant with positive sign (more on this later), and the complete absence of a certain kind of field—called a massless scalar field—that string theories predict in abundance. Thus, it seems likely that even if string theory is true in some generalized sense, the actual theory describing our universe must differ significantly from all string theories so far invented.

Another reason that string theory cannot be the final word is that in string theory one studies strings moving in a fixed classical spacetime. Thus, string theory is what we call a back-ground-dependent approach. It means that one defines the strings as moving in a fixed space and time. This may be a use-ful approximation, but it cannot be the fundamental theory. One of the fundamental discoveries of Einstein is that there *is* no fixed background. The very geometry of space and time is a dynamical system that evolves in time. The experimental ob-servations that energy leaks from binary pulsars in the form of gravitational waves—at the rate predicted by general relativity to the unprecedented accuracy of eleven decimal places—tell us that there is no more a fixed background of spacetime geometry than there are fixed crystal spheres holding the planets up. The fundamental theory must unify quantum the-ory with a completely dynamical description of space and time. It must be what we call a background-independent the-ory. Loop quantum gravity is such a one; string theory is not.

The debate between proponents of background-dependent and background-independent theories is in fact just the modern version of an ancient debate. Since the Greeks, the argument has raged between those who believed that space and time have an eternally fixed, absolute character and those who thought space and time are no more than relations between events that themselves evolve in time. Plato, Aristo-tle, and Newton were absolutists. Heraclites, Democritus, Leibniz, Mach, and Einstein were relationalists. When we demand that the quantum theory of gravity be background-independent, we are saying we believe that the triumph that general relativity represented for the relational point of view is final and will not be reversed.

Much of the argument between string and loop theorists is a continuation of this debate. Most string theorists were trained as elementary-particle physicists and worked their whole lives in a single fixed spacetime. Many of them have never even heard of the relational/absolute debate, which is the basic historical and philosophical context for Einstein's work. Most people who work in loop quantum gravity do so because at some point in their education they understood the relational, dynamical character of spacetime as described in general relativity, and they believe in it. They don't work on string theory because they cannot take seriously any candidate for a quantum theory of gravity that is background-dependent and hence loses (or at best hides) the relational, dynamical character of space and time.

Similarly, at first string theorists were resistant to the idea that the fundamental theory must be background-independent. However, I think that by now almost all string theorists have come around. They did so because there are reasons internal to string theory to believe that the fundamental theory must be background-independent. This is because string theory turned out to be non-unique. While the original hope, back in the 1980s, was that mathematical consistency would suffice to determine the unified theory, it turns out that in fact there are a huge number of equally consistent string theories. Each is as consistent as any other, and each depends on a different choice of fixed background. Further, in spite of the huge numbers of string theories we know about, none of them agree with observations on the three points I mentioned above.

As a result, in a move called "the second string revolution" in the middle 1990s, string theorists postulated that all the different string theories so far discovered, plus an infinite number of so-far-undiscovered theories, are but approximations to one unified theory. This theory has been called M theory, but there is no general agreement as to what its principles are or what mathematical form it takes. The idea is that M theory, if it exists, would be background-independent and have all the different background-dependent string theories as different solutions to it.

Many string theorists now say that the main problem in string theory is to find M theory and give string theory a background-independent form. But the funny thing is that not many string theorists have tried to work on this problem. The problem is that all their intuition and tools are based on background-dependent theories. When I bother string theorists about this, they tell me it's premature—not time to work on this problem yet.

I've had a lot of interesting conversations with the leaders of string theory—Edward Witten, Leonard Susskind, Renate Kallosh, David Gross, John Schwarz, Michael Green, Andrew Strominger, and many others. We clearly disagree about methodology. They tell me I have the wrong idea about how science works. They tell me one cannot hope to solve fundamental problems by attacking them directly. Instead one must follow the theory where it goes. A leading string theorist has said to me several times that "I learned a long time ago that string theory is smarter than I am" and that to try to tell the theory where to go would be to presume that you are "smarter than the theory." Another tells me that string theory works because it is "a very disciplined community" in which the leaders impose an order on the community of researchers to ensure that only a few problems are worked on at any one time.

I have huge respect for the string theorists as people and for what they have accomplished. Some of them are good friends. At the same time, I think they're wrong about how science works. I certainly don't want to say that I'm smarter than string theory, or than string theorists. But I disagree about the methodology, because I'm sure that fundamental scientific problems are not solved in such an accidental way.

Einstein used to complain that many scientists limit themselves to easy problems—"drilling where the wood is thin," as he put it. On one of the few occasions when I talked to Richard Feynman, he said that many theoretical physicists spend their careers asking questions that are only of mathematical interest. "If you want to discover something significant," he told me, "only work on questions whose answers will lead to new experimental predictions."

I also learned from the philosopher Paul Feyerabend the importance of conflict and pluralism in science. I read him in graduate school and I felt imediately that, unlike other philosophers I had been reading, he really understood what we scientists actually do. He pointed out that science often develops out of the tension that arises when competing research programs collide. He advised that in such situations one should always work on the weakest part of each of the competing programs. He also emphasized that pluralism in science is good, not bad. According to him, and I agree, science moves fastest when there are several healthy competing approaches to a problem, and stagnates when there is only one approach. I think this is true on every level—in the scientific community as a whole, in a research center or group, and even in each one of us.

So while I disagree with the leading string theorists about methodology, this hasn't kept me from working on string theory. After all, they don't own it; its open problems are there for anyone to try to solve. So I decided a few years ago to ignore their advice and try to construct the background-independent form of M theory. In the process of inventing loop quantum gravity, we gained a lot of knowledge about how to make quantum theories of space and time that are background-independent. We have a mathematical language, we have a conceptual language, we know what questions to ask, and we know how to do calculations. It turns out that there is a lot of loop quantum gravity that can be generalized and extended by adding extra dimensions and extra symmetries in order to make it a suitable language for M theory.

At first some of my friends and collaborators were shocked that I was working on string theory. However, I had an idea that maybe string theory and loop quantum gravity were different sides of the same theory, much like the parable of the blind men and the elephant. I spent about two years working on string and M theory, with the goal of making them background-independent and thus unifying string theory and loop quantum gravity. I did find some very interesting results. I was able to build a possible background-independent formulation of string theory.

The most interesting results I found use some beautiful mathematics, having to do with a kind of number called an octonion. These are numbers that you can divide, but they fail to satisfy the other rules, such as commutativity and associativity. Feza Gürsey, from Yale University and his students, especially Murat Gunyadin, have for years been exploring the idea that octonions might be connected to string theory. Using octonions, I was able to develop an attractive idea (from Corrine Manogue and Tevian Dray of Oregon State University) that explains why space may look three-dimensional while being, in a certain mathematical sense, nine-dimensional. I don't know if the direction I took is right, but I did find that it is indeed not so hard to use background-independent methods to formulate and study conjectures about what M theory is.

Working on string theory using the methods from loop quantum gravity was a lot of fun. I was out there with just a few friends, as it had been in the early days of loop quantum gravity, and I made real progress. However, in the last year I put this work aside because of the new experimental developments. As soon as I understood what Giovanni Amelino-Camelia was saying, I realized that this was science and that's what we had to focus on. Since then, it's been a lot harder to wake up and go to work in the morning to an imaginary world with six or seven extra dimensions.

There was another piece of shocking news from the experimenters that took me away from string theory: the discovery over the last couple of years that most of the energy in the universe is in a form that Einstein called the cosmological constant. The cosmological constant can be interpreted as indicating that empty space has a certain intrinsic energy density. This is a hard thing to believe in, but the cosmological data cannot now be explained convincingly unless one assumes that most of the energy of the universe is in this form. The problem is that string theory seems to be incompatible with a world in which a cosmological constant has a positive sign, which is what the observations indicate. This is a problem that string theorists are thinking and worrying very hard about. They are resourceful people, and maybe they'll solve it, but as things stand at the moment, string theory appears to be incompatible with that observation.

Meanwhile, loop quantum gravity incorporates a positive cosmological constant extremely well. In fact, it's our best case: If there's a cosmological constant, we're able to find a candidate for the quantum state of the universe and show that it predicts that the universe at large scales is governed by general relativity and quantum theory. So in the last several months, I've mostly been studying how to make predictions about the new experiments from a version of loop quantum gravity that incorporates a positive cosmological constant.

The good thing about science is that you get these shocks from the real world. You can live for a few years in an imaginary world, but in the end the task of science is to explain what we observe. Then you look in the mirror and ask yourself, "Do I want to be out there in eleven dimensions, playing with beautiful math, when the experiments start coming in?"

A LOOK AHEAD

MARTIN REES

The challenge is to understand how complexity emerges. This is just as fundamental as the challenge to come up with the so-called theory of everything—and it is independent of it. The theoretical physicist Steven Weinberg says that if you go on asking "Why . . . why . . . why?" you get back to a question in particle physics or cosmology. That's true to a degree, but only in a limited sense.

SIR MARTIN REES is Royal Society Professor at Cambridge University, a fellow of Kings College, and the U.K.'s Astronomer Royal. He was previously Plumian Professor of Astronomy and Experimental Philosophy at Cambridge, having been elected to this chair at the age of thirty, succeeding Fred Hoyle. He is the author of several books, including *Gravity's Fatal Attraction* (with Mitchell Begelman), *Before the Beginning, Just Six Numbers, Our Cosmic Habitat*, and *Our Final Hour.*

The problems posed by the ultra-early universe are coming into focus. We now know the key properties of the universe at the present era—its density, its age, and its main constituents. Indeed, the last few years will go down as especially remarkable in the annals of cosmology, because within those years we've pinned down the shape and contents of the cosmos, just as in earlier centuries the pioneer navigators determined the size of the earth and the layout of its continents. The challenge now is to explain how the universe got that way, to understand why the universe is expanding the way it is, and why it ended up with the content it has. We can trace its history back to about a microsecond after the putative Big Bang that started it off, but what happened in that first, formative microsecond? The boisterous variety of ideas being discussed—branes, inflation, and so on—makes it clear that we're still a long way from the right answer. We're at the stage where all possibilities should be explored. It's worthwhile to consider the consequences of even the most flaky ideas, although the chance of any of them panning out in the long run is not very high.

I wouldn't claim to be a technical expert in any of the specific theories for the ultra-early universe. It seems likely that extra dimensions of space are going to play a role. Moreover, the idea of inflation, which has dominated the field for twenty years, is now being generalized by other concepts from people like Lisa Randall, Neil Turok, and Paul Steinhardt. The key goal, of course, is to develop a convincing all-encompassing theory describing the early universe and making testable predictions about the world today. If we had a theory that gave us a deeper and more specific understanding of the masses of electrons and protons and the forces governing them than the so-called standard model does, we'd take seriously its implications for the ultra-early universe. The hope is that one of the exotic new theories will make testable predictions either about the ordinary world of particles or about the universe. Some of them, for instance, make distinctive predictions about the amount of gravitational radiation filling the universe. We can't measure this today, but within ten years we might be able to. That's one way in which astronomical observations might narrow the range of options.

The easiest idea to understand conceptually is eternal inflation, which Alan Guth advocates and on which the Stanford cosmologist Andrei Linde has done a great deal of detail work. This naturally gives rise to many Big Bangs. Whether those Big Bangs will be close replicas of each other or whether the material in each of them would be governed by different laws is something we don't know. Eternal inflation may bypass the complications of extra

dimensions and quantum gravity because these are rele-
gated to the infinite past.

Most of us, however, suspect that a prerequisite for
progress will be a worked-out theory that relates gravity to
the microworld. Back at the very beginning, the entire universe could have been squeezed to the size of an elementary
particle; quantum fluctuations could have shaken the entire
universe, and there would be an essential link between cosmology and the microworld. String theory and M theory—
both involving extra dimensions—are the most ambitious
and currently fashionable attempts to do that. When we have
that theory, we at least ought to be able to formulate some
physics for the very beginning of the universe. One question,
of course, is whether we will find that space and time are tan-
gled in such a convoluted way that we can't really talk about
a "beginning" in time. We will have to jettison more and
more of our common-sense concepts as we go to these extreme conditions. The main stumbling block at the moment
is that the mathematics involved in these theories is so difficult that it's not possible to relate the complexity of this
ten- or eleven-dimensional space to anything we can actually
observe. In addition, although these theories appear aestheti-
cally attractive and give us a natural interpretation of gravity,
they don't yet tell us why our three-dimensional world contains the types of particles that physicists study.

Although Roger Penrose can probably manage four
dimensions, I don't think any of these theorists can in any
intuitive way imagine several extra dimensions. They can,
however, envision them as mathematical constructs, and
certainly the mathematics can be written down and studied.
The one thing that is rather unusual about string theory—from
the viewpoint of the sociology and history of science—is that
it's one of the few instances in which physics has been held

up by a lack of the relevant mathematics. In the past, physicists generally took fairly old-fashioned mathematics off the shelf. Einstein used nineteenth-century non-Euclidean geometry, and the pioneers in quantum theory used group theory and differential equations that had essentially been worked out long beforehand. But string theory poses mathematical problems that aren't yet solved and has actually brought math and physics closer together. It's the dominant approach right now, and it has had some successes already, but the question is whether it will develop to the stage where we can solve problems that can be tested observationally. If we can't bridge the gap between this ten-dimensional theory and anything we can observe, it will grind to a halt.

Our models of the ultra-early universe are today rather like the generic Big Bang model in the decades before the 1960s, when people like Georges Lemaître, George Gamow, and Alexandr Friedman formulated some basic ideas even though no one could really test them, and the physics of the first few minutes was still entirely conjectural. In the same way, inflation and string theories of the ultra-early universe are ahead of any testable predictions. The question is whether in ten or twenty years we will have ways of testing them, just as for the last ten years we have had very good tests of the Big Bang theory back to the stage when the universe was a second old. If these ideas can never be tested, then of course one could argue that they are no more than "ironic science," in the disparaging sense of that phrase introduced by John Horgan. But I hope that within ten or twenty years we'll know which, if any, of them is on the right track—either because one of them will be part of a general unified theory explaining the basic forces and laws of nature or because some astronomical observation capable of discriminating between them will have been made. Once again, theorists are leading, goading, and stimulating the observers and experimenters. It's important that alternatives to mainstream ideas are being explored—for example, Lee Smolin's work on loop quantum gravity. The one thing that concerns me about string theory is the perhaps excessive concentration of talents in that field. It's not only a suboptimal deployment of scientific effort, but it's sure to lead to a lot of disillusionment when so many brilliant young people are all chasing the same ideas.

I'm also interested in some fundamental questions about the uniqueness of physical laws. I've always been impressed by so-called fine-tuning arguments—arguments holding that our universe seems to be rather special, and its complexity arose only because its laws have a highly unusual character. Our existence is a genuine mystery, since you can easily imagine a set of laws that would lead to a sterile or a stillborn universe. The most natural answer to the mystery would be that our Big Bang wasn't the only one—that there are many universes, and they have ended up being governed by different laws, only some of which allow structures, and eventually life, to evolve. So I'm attracted to the cosmological models that allow not just one Big Bang but many—one feature of the eternal-inflation scenario pioneered by Andre Linde and of some of the models calling for extra dimensions. I'd like to know whether the (still speculative) physics that predicts these multiple universes is correct, and whether the various universes are governed by different physical laws and different forces. Do they contain particles wholly unlike the particles that make up our own universe? If there is indeed a huge variety among the many universes, then it should occasion no surprise if there were at least one universe of the kind we inhabit.

Another perspective comes from the Oxford theoretician David Deutsch, who has refined the so-called "many worlds" theory of quantum mechanics. He's thinking of these universes as being somehow superimposed on each other—which is not the same idea as Lisa Randall's parallel universes. A clearer understanding of quantum theory and quantum computation may be arrived at by thinking along those lines. There's truth in John Polkinghorne's remark that "Your average quantum mechanic is no more philosophical than your average motor mechanic." Most physicists just use the theory, in a rather mindless way, to solve probems. Quantum mechanics might give you the answers, but there are still mysteries about it, and we shouldn't assume we've got the right way of looking at it yet. People like David Deutsch are heading us in a productive direction.

There's a tendency to use terms like "Theory of Everything" and "Final Theory" to denote what theorists like Edward Witten of the Institute for Advanced Study—and hundreds of other talented theorists—are seeking. The theory they're looking for would be the end of a quest that started with Newton and continued through Einstein and his successors. But of course it wouldn't be the end of science, it would just be the end of a particular quest. It wouldn't help us to understand most of the complexities of the world. Most scientists, even most physicists, wouldn't be helped at all by a fundamental theory, because the difficulties that confront them are not the result of not knowing the basic laws. The challenge is to understand how complexity emerges. This is just as fundamental as the challenge to come up with the so-called theory of everything—and it is independent of it. The theoretical physicist Steven Weinberg says that if you go on asking "Why . . . why . . . why?" you get back to a question in particle physics or cosmology. That's true to a degree, but only in a limited sense. It's a challenge to ask why a fluid sometimes behaves in a regular way and sometimes in a chaotic way—to understand turbulence, or dripping taps, for example—but the answer won't come from analyzing the liquid down to its subatomic constituents. It will come by thinking in a quite different way, about complexity. To give an example, Mitchell Feigenbaum's discovery that the same series of numbers come up in the transition from ordered to chaotic behavior is an important discovery about the world, but it's got absolutely nothing to do with particle physics, even though it is just as fundamental.

Before the age of computers, you could never fully appreciate how a simple algorithm could result in tremendous complexity. It's through the computer that we've been able to do this new kind of science—of which, of course, Stephen Wolfram is the highest-profile propagandist—which allows us to develop new intuitions about how simple patterns and simple algorithms can have extremely complex consequences. This is a science fully on the level of particle physics and string theory intellectually but quite disjoined from them. Wolfram has produced a very fine manifesto for this kind of science, but whether his way of looking at things is the key to understanding space, time, and particles, I don't know. I'm rather skeptical about it, to be honest. But I'm sympathetic to people who, like Princeton physicist Philip Anderson, want to deflate the hubris of the fundamental physicists who claim that their subject is the deepest and highest priority of all. It's just as important to understand complexity, to see it in the simplest form in the transition to chaos and in more complicated forms in all the rest of science: the genetic code, fluid flows, and all the rest of it.

My own preoccupation right now—as indeed it has been for more than twenty years—is to understand how the cosmic dark age ended. After the initial brilliance of the Big Bang, the universe cooled and darkened until it was lit up again when the first stars or galaxies formed. We're making great progress, with the aid of both observation and theory, in understanding how the universe went from being amorphous and structureless to becoming complex. This key transition happened quite late—perhaps 100 million years after the Big Bang. The basic physics at the prevailing low densities and temperatures is uncontroversial, but things get complicated for the same reason that all environmental science is. I'm trying to understand how the first structures evolved—how the first stars, black holes, and galaxies developed.

But cosmologists like myself are no less concerned than anyone else with what happens next week or next year; indeed, their awareness of the vast eons that stretch ahead perhaps makes them especially mindful of life's future—its posthuman potential. There's a lot of jubilation about the accelerating progress of certain sciences, and of course there are people, like Ray Kurzweil in particular, who think that technical progress is running away toward some kind of singularity, or cusp, that could be reached in about fifty years. My concern is that these advances—particularly advances in biotechnology—will lead to greater instability. They increase the leverage and power of a single disaffected person or a small group. It will take only a few people, with the tremendous leverage that technology offers, to cause disasters that could disrupt our whole society. Especially if everyone knows that such disasters could repeat anytime and cannot feasibly be prevented. The anthrax episode in 2001 exemplifies that even a well-contained outbreak of that kind can affect the psyche of a whole society. The media and the general hype can amplify any scare, because we're so connected, so networked. I can't see how we can avoid having episodes that completely seize up society—or even cause it to collapse. I'm pessimistic, because it seems to me that it will be very hard to guard against such things. Twenty years ago, we worried about a possible confrontation between superpowers; in the 1990s, we worried about rising nationalism and smaller-scale conflicts. Now we worry about terrorists and other disaffected groups, and in the future we'll have to worry about disaffected individuals with the mindset of those who now design computer viruses but will soon have the power to do far worse.

Such thoughts make me depressed about what's going to happen in the next ten or twenty years. If we can stave off disaster, however, then I'm with Kurzweil in expecting that the rate of change in our life will be faster in the coming fifty years than it was in the last fifty.

SUGGESTED READING

JOHN BROCKMAN
The Next Fifty Years: Science in the First Half of the Twenty-First Century (Vintage, 2002)
The Third Culture: Beyond the Scientific Revolution (Simon & Schuster, 1995)

RODNEY BROOKS
Flesh and Machines: How Robots Will Change Us (Pantheon Books, 2002)

ANDY CLARK
Natural-Born Cyborgs: Why Minds and Technologies Are Made to Merge (Oxford University Press, 2003)
Being There: Putting Brain, Body, and World Together Again (MIT Press, 1997)

HELENA CRONIN
The Ant and the Peacock: Altruism and Sexual Selection from Darwin to Today (Cambridge University Press, 1992)
Breaking the Spell: Religion as a Natural Phenomenon (Viking, 2006)

DANIEL C. DENNETT
Freedom Evolves (Viking, 2003)
Kinds of Minds: Toward an Understanding of Consciousness (Basic Books, Science Masters series, 1996)
Darwin's Dangerous Idea: Evolution and the Meanings of Life (Simon & Schuster, 1995)
Consciousness Explained (Little, Brown, 1991)

DAVID DEUTSCH
The Fabric of Reality: The Science of Parallel Universes—and Its Implications (Penguin USA, 1998)

JARED DIAMOND

Collapse: How Societies Choose to Fail or Succeed (Viking, 2004)

Guns, Germs, and Steel: The Fates of Human Societies (W. W. Norton, 1999)

Why Is Sex Fun? The Evolution of Human Sexuality (Basic Books, Science Masters series, 1997)

The Third Chimpanzee: The Evolution and Future of the Human Animal (HarperCollins, 1992)

DAVID GELERNTER

1939: The Lost World of the Fair (Free Press, 1999)

Machine Beauty: Elegance and the Heart of Technology (Basic Books, Master Minds series, 1998)

Drawing a Life: Surviving the Unabomber (Free Press, 1997)

The Muse in the Machine: Computerizing the Poetry of Human Thought (Free Press, 1994)

Mirror Worlds: Or the Day Software Puts the Universe in a Shoebox— How It Will Happen and What It Will Mean (Oxford University Press, 1991)

ALAN GUTH

The Inflationary Universe: The Quest for a New Theory of Cosmic Origins (Perseus, 1997)

MARC D. HAUSER

Moral Minds: How Nature Designed our Universal Sense of Right and Wrong (Ecco, 2006)

Wild Minds: What Animals Really Think (Henry Holt, 2000)

STEPHEN M. KOSSLYN

Graph Design for the Eye and Mind (Oxford University Press, 2006)

Psychology: The Brain, the Person, the World (with Robin S. Rosenberg) (Allyn & Bacon, 2000)

Image and Brain: The Resolution of the Imagery Debate (MIT Press, 1994)

Wet Mind: The New Cognitive Neuroscience (Free Press, 1992)

RAY KURZWEIL

The Singularity is Near: When Humans Transcend Biology (Viking, 2006)

The Age of Spiritual Machines: When Computers Exceed Human Intelligence (Viking, 1999)
The Age of Intelligent Machines (MIT Press, 1992)

JOSEPH LEDOUX
Synaptic Self: How Our Brains Become Who We Are (Viking, 2002)
The Emotional Brain: The Mysterious Underpinnings of Emotional Life (Simon & Schuster, 1996)

MARVIN MINSKY
The Emotion Machine: Commonsense Thinking, Artificial Intelligence, and the Future of the Human Mind (Simon & Schuster, 2006)
The Society of Mind (Simon & Schuster, 1987)

HANS MORAVEC
Robot: Mere Machine to Transcendent Mind (Oxford University Press, 1998)
Mind Children: The Future of Robot and Human Intelligence (Harvard University Press, 1988)

STEVEN PINKER
The Stuff of Thought: Language as a Window into Human Nature (Viking, 2007)
The Blank Slate: The Modern Denial of Human Nature (Viking, 2002)
Words and Rules: The Ingredients of Language (Basic Books, Science Masters series, 1999)
How the Mind Works (W. W. Norton, 1997)
The Language Instinct: How the Mind Creates Language (William Morrow, 1994)

MARTIN REES
Our Final Hour: A Scientist's Warning: How Terror, Error, and Environmental Disaster Threaten Humankind's Future In This Century—On Earth and Beyond (Basic Books, 2003)
Our Cosmic Habitat (Princeton University Press, 2001)
Just Six Numbers: The Deep Forces That Shape the Universe (Basic Books, Science Masters series, 1999)
Before the Beginning: Our Universe and Others (Perseus, 1997)

LEE SMOLIN

The Trouble With Physics: The Rise of String Theory, The Fall of a Science, and What Comes Next (Houghton Mifflin, 2006)

Three Roads to Quantum Gravity (Basic Books, Science Masters series, 2001)

The Life of the Cosmos (Oxford University Press, 1997)

RICHARD WRANGHAM

Demonic Males: Apes and the Origins of Human Violence (with Dale Peterson) (Houghton Mifflin, 1996)

ACKNOWLEDGMENTS

From the very beginning of *Edge*, I have received a great deal of encouragement and support from key people in the Barnes & Noble organization, including Steve Riggio, Mike Ferrari, and Michael Friedman. They approached me with the idea that a volume based on *Edge* (www.edge.org) would make a valuable book, and I thank them for the suggestion and their encouragement. I also wish to thank Michael Fragnito and Laura Nolan of Barnes & Noble Publishing for backing this project.

Russell Weinberger, associate publisher of *Edge*, has been involved in all aspects of publication and Christopher Williams worked closely with me editorially on the initial transformation of many of the Q&A transcripts into essay form, as well as providing English translations of German texts. I wish to thank them both for their valuable contributions.

I wish to thank Judy Herrick of Typro for her work in transcribing all the interviews. And finally, I am indebted to Sara Lippincott for her thoughtful and meticulous editing.

INDEX